AF564411

Aquaculture Nutrition

NIPA® GENX ELECTRONIC RESOURCES & SOLUTIONS P. LTD.
New Delhi-110 034

About the Editor

Dr. Ishtiyaq Ahmad earned his doctorate from the Fish Nutrition Research Laboratory, Department of Zoology, University of Kashmir, Srinagar, India. He is currently a Senior Researcher at the Division of Fish Genetics and Biotechnology, Faculty of Fisheries, Sher-e-Kashmir University of Agricultural Sciences and Technology Kashmir (SKUAST-K), Rangil, Ganderbal, India. His research focuses on developing pure lines for the genetic improvement of rainbow trout to enhance overall aquaculture production. Dr. Ahmad's expertise spans Fish Nutrition, Fish Physiology and Biochemistry, Nutrigenomics, Fish Parasitology, and Molecular Biology. His notable research includes establishing the optimal branched-chain amino acid requirements for rainbow trout. He has published over 40 research articles in esteemed international journals such as *Scientific Reports, Aquaculture, Reviews in Aquaculture, Aquaculture Nutrition, Aquaculture Research, Aquaculture Reports, Journal of Applied Ichthyology, Frontiers in Marine Science, Food Chemistry: X,* and *Journal of Contaminant Hydrology*. Additionally, Dr. Ahmad has authored several book chapters and five books with national and international publishers. He has presented his work at various state, national, and international conferences, seminars, and workshops, earning numerous accolades, including the Young Scientist Award from a national organization and several Best Paper Presentation awards. Dr. Ahmad serves on the editorial boards of prominent international journals, including *PLOS ONE* and *Frontiers in Nutrition*. He also worked as a Senior Research Fellow in a DBT-funded project on the nutrient requirements of rainbow trout fingerlings and is a recipient of the prestigious CSIR NET award.

Aquaculture Nutrition
Nanotechnology, Sustainable Feed Innovations and Precision Feeding Strategies

Ishtiyaq Ahmad, Ph.D
Senior Researcher
Division of Fish Genetics and Biotechnology
Faculty of Fisheries, She-r-e Kashmir University
of Agricultural Science and Technology
Kashmir, Ganderbal, India-190025

NIPA® GENX ELECTRONIC RESOURCES & SOLUTIONS P. LTD.
New Delhi-110 034

NIPA® GENX ELECTRONIC RESOURCES & SOLUTIONS P. LTD.

101,103, Vikas Surya Plaza, CU Block
L.S.C. Market, Pitam Pura, New Delhi-110 034
Ph : +91-11-43860225, Mob.: +91 9717133558, 9540816132
E-mail: newindiapublishingagency@gmail.com
Website: www.nipaersources.com

Print ISBN: 978-93-58872-38-5
ebook ISBN: 978-93-58879-62-9

NIPA® also publishes books in a variety of electronic formats. Some content that appears in print may not be available in electronic books, and vice versa.

Composed and Designed by NIPA®.

Preface

In recent years, aquaculture has emerged as a vital component of global food security, supplying nutritious seafood to millions worldwide. However, as the industry grows, it faces pressing challenges that demand forward-thinking solutions-among them, the need for sustainable and efficient nutrition practices. **Aquaculture Nutrition: Nanotechnology, Sustainable Feed Innovations and Precision Feeding Strategies** explores a range of breakthroughs that are transforming how aquaculture nutrition is approached, with the aim of fostering sustainability, enhancing feed efficiency, and promoting the health and welfare of aquatic species.

This book presents a holistic overview of contemporary advancements in aquaculture nutrition, focusing on three transformative domains: the application of nanotechnology, the development of sustainable feed ingredients, and the implementation of precision feeding strategies. Nanotechnology is breaking new ground in aquaculture, offering innovative solutions for nutrient delivery, water quality monitoring, and disease control. By leveraging nanoscale technology, this field is enabling targeted and efficient nutrient uptake, potentially reducing feed costs and minimizing environmental impact.

Sustainable feed innovation is a central theme, addressing the need to move away from over-reliance on traditional marine resources such as fishmeal and fish oil. Alternative protein and lipid sources-including insect-based proteins, algae, and plant-derived ingredients-are investigated for their potential to reduce the ecological footprint of aquafeeds while supporting optimal growth and health outcomes in farmed species. Additionally, functional additives derived from natural sources are discussed for their roles in enhancing immune function, growth rates, and resistance to disease, contributing to a more resilient aquaculture sector.

Precision feeding strategies represent the third major area of advancement, introducing techniques that align with the specific nutritional needs of fish across their life stages. By integrating emerging technologies such as artificial intelligence, machine learning, and sensor-based monitoring, precision feeding allows for real-time adjustments that improve feed conversion ratios and reduce waste. These strategies not only enhance production efficiency but also support environmental sustainability, aligning with industry goals of responsible resource use.

Aquaculture Nutrition is intended as a comprehensive resource for researchers, nutritionists, students, and industry practitioners seeking to stay informed of the latest innovations in this dynamic field. By drawing on recent research and practical applications, this book aims to inspire new pathways toward sustainable and efficient aquaculture nutrition practices. I hope this book will serve as a foundation for continued progress and will motivate ongoing research and development to meet the nutritional challenges facing aquaculture today.

Ishtiyaq Ahmad

Acknowledgement

The journey of creating **Aquaculture Nutrition: Nanotechnology, Sustainable Feed Innovations, and Precision Feeding Strategies** has been both inspiring and rewarding, and it would not have been possible without the support, insights, and dedication of numerous individuals and institutions. I extend my heartfelt gratitude to each contributor who has played a part in shaping this work.

First and foremost, I would like to thank the researchers, scientists, and industry experts who shared their cutting-edge findings and valuable perspectives on the transformative innovations within aquaculture nutrition. Their commitment to advancing sustainable practices has been instrumental in addressing the challenges faced by the aquaculture sector today. Each chapter in this book represents the collective expertise and visionary approaches of individuals who are shaping the future of aquaculture.

I am deeply grateful to my colleagues and mentors, whose encouragement and guidance have been invaluable throughout the process. Their insights have provided clarity, direction, and inspiration as I worked to compile, organize, and edit this volume. Their unwavering belief in the importance of this project has driven me to delve deeply into each topic, ensuring it reflects both the latest advancements and the industry's aspirations for sustainable growth.

A special thanks goes to the publishing team, whose professionalism, patience, and dedication have been essential in bringing this book to fruition. From editorial support to production, their meticulous attention to detail has been instrumental in transforming this vision into a polished, accessible resource. I am also grateful to the peer reviewers for their constructive feedback, which has strengthened the quality of each chapter and ensured the book meets the highest academic standards.

Finally, I wish to thank my family and friends for their understanding, encouragement, and support throughout this endeavor. Their patience and unwavering belief in me have been a constant source of motivation.

Aquaculture Nutrition is the result of collective effort and shared dedication to advancing sustainable aquaculture practices. It is my sincere hope that this book will serve as a valuable resource for researchers, students, and professionals in the field, inspiring further innovation and progress in aquaculture nutrition.

Ishtiyaq Ahma

Contents

1

Nanotechnology in Aquafeeds Enhancing Nutrient Delivery

1. Introduction

Aquaculture is rapidly emerging as one of the most important sectors in global food production, supplying an increasing share of the world's demand for high-quality protein. As global populations rise and overfishing depletes wild fish stocks, aquaculture has become essential in addressing the global food security challenge. The expansion of aquaculture, however, presents a new set of challenges, particularly in the formulation and delivery of efficient and sustainable aquafeeds. Fish and crustaceans, like all living organisms, require precise nutrition for optimal growth, immune function, and reproduction. In commercial aquaculture, ensuring that farmed species receive the right balance of nutrients through formulated feeds is crucial to achieving high productivity and minimizing environmental impact.

`Traditional aquafeeds are designed to meet the nutritional needs of farmed species; however, they face several obstacles when it comes to efficient nutrient delivery. The aquatic environment poses unique challenges for feed stability and nutrient bioavailability, including:

- **Water Stability:** Aquatic feeds must resist rapid breakdown when exposed to water, ensuring that nutrients remain intact until they are ingested by fish or crustaceans.
- **Nutrient Leaching:** In water, nutrients can leach out of feeds before they are consumed, reducing their availability and increasing waste.
- **Complex Dietary Requirements:** Aquatic species have diverse and often complex dietary needs, which must be met through balanced and targeted feed formulations.

To address these challenges, researchers and feed manufacturers are exploring advanced technologies, such as nanotechnology, to improve nutrient delivery in aquafeeds. Nanotechnology offers unique opportunities to enhance the nutritional value and efficiency of aquafeeds through the development of nanostructured materials, nanoparticles, and encapsulation techniques.

These technologies can enhance nutrient bioavailability, stabilize sensitive compounds, and provide controlled release mechanisms that optimize the uptake of nutrients by aquatic species.

This chapter will delve into the application of nanotechnology in aquafeeds, discussing how it enhances nutrient delivery, improves feed efficiency, and supports the health and growth of farmed aquatic species. The focus will be on how specific nanotechnology tools—such as nanoparticles, nanoencapsulation, and nanostructured carriers—can address the inherent limitations of traditional aquafeeds, while also contributing to sustainability and environmental stewardship in aquaculture.

2. Overview of Nanotechnology in Aquaculture

Nanotechnology refers to the science and engineering of materials at the nanoscale, typically between 1 and 100 nanometers in size. At this small scale, materials often exhibit unique physical, chemical, and biological properties compared to their larger, bulk counterparts. These nanoscale properties can be harnessed to develop new materials and technologies with enhanced performance, making nanotechnology a promising tool for solving critical problems in diverse fields, including aquaculture.

In the context of aquafeeds, nanotechnology offers solutions that improve nutrient retention, delivery, and absorption by aquatic animals. Due to the small size and large surface area of nanomaterials, they can interact more effectively with biological systems, enhancing the bioavailability of nutrients and promoting better uptake. Additionally, nanostructures can be engineered to protect sensitive nutrients from environmental degradation, control the release of nutrients, and increase the stability of functional feed additives.

Nanotechnology in aquaculture has several primary applications:

1. **Nanoparticles:** These are small particles, often created from minerals, proteins, or lipids, that are used to improve the bioavailability of nutrients and enhance the stability of feeds in water. Nanoparticles have unique physicochemical properties that allow them to interact more effectively with cellular membranes, leading to improved nutrient absorption.
2. **Nanoencapsulation:** This is a technique where nutrients or bioactive compounds are encapsulated within nanostructured carriers, such as nanoliposomes or polymer-based nanoparticles. Nanoencapsulation protects these compounds from oxidation, degradation, and leaching, ensuring their delivery to the target site within the digestive system of aquatic species. This method also allows for the controlled and sustained release of nutrients, providing continuous nutrient supply and reducing feed waste.

3. **Nanostructured Carriers:** Nanostructured materials, such as nanofibers, nanocapsules, and nanoemulsions, are used as carriers for poorly soluble nutrients, such as lipids, vitamins, and trace minerals. These carriers improve the solubility and absorption of these nutrients in water and in the digestive tract of aquatic animals, ensuring that they are efficiently utilized for growth and health.

2.1. Key Advantages of Nanotechnology in Aquafeeds

Nanotechnology offers several advantages that make it highly suitable for use in aquafeeds, including:

- **Increased Nutrient Bioavailability:** Nutrients delivered through nanoparticles or nanoencapsulated forms have higher bioavailability due to their small size and increased surface area. This allows for better absorption by the digestive systems of aquatic animals, leading to improved growth rates and feed efficiency.
- **Controlled Release of Nutrients:** Nanotechnology can be used to design feed additives that release nutrients in a controlled manner, ensuring a continuous supply of essential nutrients over time. This reduces the risk of nutrient deficiencies and enhances the overall health of farmed fish and crustaceans.
- **Protection of Sensitive Nutrients:** Many essential nutrients, such as vitamins and fatty acids, are prone to degradation due to heat, light, and oxygen exposure. Nanotechnology can encapsulate these sensitive compounds in protective nanocarriers, safeguarding them during feed processing, storage, and water exposure.
- **Reduced Nutrient Leaching:** By using nanoparticles or nanostructured carriers, nutrients can be retained in the feed for longer periods, preventing their rapid dissolution and leaching into the water. This improves nutrient utilization and reduces waste, contributing to more sustainable aquaculture practices.
- **Improved Feed Stability in Water:** Traditional aquafeeds often break down quickly when immersed in water, leading to nutrient loss and decreased feeding efficiency. Nanotechnology-based feeds can improve water stability, ensuring that feeds remain intact until consumed by aquatic animals.

2.2. Addressing Environmental Sustainability in Aquafeeds

One of the significant concerns in aquaculture is the environmental impact of excess nutrients, particularly from uneaten feed and nutrient leaching, which can contribute to water pollution, eutrophication, and habitat degradation. By enhancing the efficiency of nutrient delivery and reducing nutrient waste, nanotechnology has the potential to make aquaculture more sustainable and environmentally friendly. Key ways in which nanotechnology supports sustainability in aquafeeds include:

- **Minimized Feed Waste:** Nanotechnology improves nutrient retention and absorption, leading to less uneaten feed and nutrient excretion into the environment. This helps reduce the environmental footprint of aquaculture operations.
- **Better Resource Utilization:** Nanotechnology enables more efficient use of resources, such as fishmeal and fish oil, by increasing the bioavailability of essential nutrients. This reduces the need for high inclusion levels of these ingredients in aquafeeds and promotes the development of alternative, sustainable feed sources.
- **Reduced Dependency on Antibiotics and Chemicals:** By enhancing the immune function and overall health of farmed species, nanotechnology-based feeds can reduce the reliance on antibiotics, chemicals, and other medications, contributing to more sustainable and environmentally friendly aquaculture practices.

3. Nanoparticles in Aquafeeds

Nanoparticles, due to their small size and unique properties, can improve nutrient bioavailability and enhance feed efficiency in aquaculture. They can be used to deliver essential minerals, vitamins, lipids, and other nutrients in a more efficient manner, ensuring that aquatic animals receive the necessary nutrients for optimal growth and health. Nanoparticles in aquafeeds include nano-sized minerals, such as zinc, selenium, and calcium, which are critical for various physiological processes in fish and crustaceans. By delivering these nutrients in nanoparticle form, their absorption is improved, reducing the amount of feed needed and enhancing overall feed efficiency.

4. Nanoencapsulation for Nutrient Protection

Nanoencapsulation is another key application of nanotechnology in aquafeeds, allowing for the protection and controlled release of sensitive nutrients. This process ensures that essential nutrients, such as omega-3 fatty acids, vitamins, and probiotics, remain stable during feed storage and delivery, maximizing their effectiveness in supporting the health and growth of aquatic animals.

Nanotechnology presents an exciting frontier in aquafeeds, offering innovative solutions to the challenges of nutrient delivery, feed stability, and environmental sustainability in aquaculture. By enhancing the bioavailability of nutrients, protecting sensitive compounds, and reducing nutrient leaching, nanotechnology can significantly improve feed efficiency and contribute to the sustainable growth of the aquaculture industry. As research and development in this field continue, nanotechnology will likely play an increasingly important role in meeting the global demand for high-quality seafood while minimizing the environmental impact of aquaculture practices.

3. Nutrient Delivery Challenges in Aquafeeds

Efficient nutrient delivery in aquafeeds is critical for ensuring the optimal growth, health, and productivity of aquatic species such as fish, crustaceans, and mollusks. Aquafeeds need to provide all essential nutrients, including proteins, lipids, vitamins, minerals, and functional additives, to meet the specific dietary requirements of different species. However, several challenges exist in traditional aquafeeds due to the nature of aquatic environments and the dietary needs of aquatic animals. These challenges hinder the effectiveness of nutrient delivery, impacting both the growth performance of farmed species and the environmental sustainability of aquaculture operations. The key challenges include:

3.1. Nutrient Loss in Aquatic Environments

Aquatic environments pose a unique challenge in maintaining nutrient integrity within feeds. Once submerged, traditional pelleted or extruded feeds can quickly absorb water, leading to the leaching of water-soluble nutrients such as vitamins and minerals before they are consumed by fish or other aquatic animals. This nutrient loss not only reduces the nutritional value of the feed but also increases feed waste, making it difficult to meet the specific dietary needs of aquatic species. Moreover, feed degradation in water increases the demand for higher feed inputs, leading to inefficiencies in feed conversion ratios. This also raises production costs and contributes to nutrient pollution in aquaculture systems, affecting water quality and potentially harming aquatic ecosystems.

3.2. Poor Bioavailability of Certain Nutrients

Some nutrients, particularly lipids (fats and oils), vitamins, and minerals, have poor bioavailability due to their low solubility in water and limited ability to be absorbed by the gastrointestinal tract of aquatic species. This low bioavailability is particularly problematic for essential nutrients such as omega-3 fatty acids (EPA and DHA), fat-soluble vitamins (e.g., vitamins A, D, E, and K), and trace minerals (e.g., zinc, selenium, and iron), which play vital roles in fish growth, immune function, reproduction, and overall health. Traditional feed formulations may not fully address the issue of bioavailability, leading to suboptimal nutrient absorption and, in turn, lower growth performance and resistance to diseases in farmed species. As a result, more feed may be required to achieve the desired levels of nutrient uptake, increasing feed costs and feed conversion ratios (FCRs).

3.3. Environmental Impact of Nutrient Waste

Unused or leached nutrients from aquafeeds not only represent a waste of resources but also contribute to environmental issues such as water pollution and eutrophication. Eutrophication occurs when excess nutrients, particularly nitrogen and phosphorus, accumulate in water bodies, leading to excessive algal blooms, decreased oxygen levels, and overall degradation of water quality. This can negatively affect both farmed species and surrounding aquatic ecosystems. The environmental impact of nutrient waste from aquaculture highlights the need for more efficient feed formulations that minimize nutrient loss, improve bioavailability, and promote better resource utilization. Nanotechnology provides novel solutions to these challenges by improving the stability, protection, and bioavailability of nutrients, ultimately enhancing the efficacy of aquafeeds and minimizing waste.

4. Nanotechnology Applications in Aquafeeds

Nanotechnology offers innovative solutions to improve the delivery, absorption, and efficiency of nutrients in aquafeeds, addressing the challenges of nutrient loss, poor bioavailability, and environmental impact. By utilizing nanoparticles, nanoencapsulation, and nanostructured carriers, aquafeeds can be designed to optimize nutrient uptake, enhance growth performance, and improve fish health. Some of the key nanotechnology applications in aquafeeds are outlined below.

4.1. Nanoencapsulation of Nutrients

Nanoencapsulation involves enclosing nutrients, additives, or bioactive compounds in nanoscale carriers, such as liposomes, nanospheres, or polymer-based nanoparticles. This technique offers several advantages for nutrient delivery in aquafeeds:

- **Protection from Degradation:** Many essential nutrients, such as vitamins, probiotics, antioxidants, and omega-3 fatty acids, are sensitive to environmental factors such as light, heat, and oxygen, which can degrade their effectiveness during feed processing and storage. Nanoencapsulation protects these sensitive nutrients by creating a protective barrier around them, ensuring they remain intact and functional until they are consumed by aquatic species.

- **Controlled Release:** Nanoencapsulation allows for the controlled release of nutrients within the digestive tract of aquatic animals, providing sustained availability of nutrients over time. This controlled release mechanism ensures that nutrients are absorbed gradually, improving nutrient utilization and reducing waste. For example, nanoencapsulated vitamins and minerals can be released slowly, enhancing their absorption and bioavailability.
- **Targeted Delivery:** Nanoencapsulation can enable the targeted delivery of specific nutrients to particular organs or tissues, improving the health and growth performance of aquatic species. By directing nutrients to where they are needed most, nanoencapsulation can enhance immune function, stress resistance, and overall well-being.

For example, studies have shown that nanoencapsulated vitamins and minerals improve growth rates, immune response, and feed efficiency in fish, while minimizing nutrient loss and environmental impact.

4.2. Nanoparticles for Improved Mineral Bioavailability

Minerals, such as zinc, selenium, and iron, are essential for various physiological functions in fish, including enzyme function, immune response, and reproductive health. However, many minerals are poorly soluble in water and have low bioavailability when included in traditional aquafeeds. Nanotechnology can overcome this challenge by producing mineral nanoparticles, which have a higher surface area-to-volume ratio and improved solubility compared to their bulk counterparts.

- **Zinc Nanoparticles:** Zinc is critical for enzyme activity, immune response, and skin health in fish. Nanoparticles of zinc improve its bioavailability, leading to enhanced growth performance, immune function, and resistance to infections. Studies have shown that nano-zinc supplements result in better absorption and utilization of zinc, contributing to improved fish health.
- **Selenium Nanoparticles:** Selenium is an essential trace element with antioxidant properties, helping protect cells from oxidative stress. Nano-selenium supplements have been found to improve growth, feed conversion ratios, and antioxidant activity in fish, promoting better overall health and reducing susceptibility to diseases.

4.3. Nanocarriers for Fatty Acids and Lipid-Soluble Nutrients

Omega-3 fatty acids (EPA and DHA) and other lipid-soluble nutrients, such as fat-soluble vitamins (A, D, E, and K), are crucial for fish health, immune function, and reproductive success. However, their poor water solubility and susceptibility to oxidation make it challenging to deliver these nutrients effectively in traditional aquafeeds. Nanocarriers, such as nanoliposomes and nanoemulsions, offer a solution to these challenges by improving the stability, dispersibility, and absorption of fatty acids and lipid-soluble nutrients. Nanocarriers protect these nutrients from oxidation, preserve their integrity, and enhance their solubility in water, making them more bioavailable to fish. This leads to improved nutrient uptake, better growth performance, and enhanced fish health.

4.4. Nano-Feed Additives for Health and Immunity

Nanotechnology can also enhance the efficacy of functional feed additives, such as probiotics, prebiotics, and immunostimulants, which play a key role in promoting fish health and disease resistance.

- **Nano-Probiotics:** Probiotics are beneficial microorganisms that improve gut health and enhance the immune response in fish. Encapsulating probiotics in nanocarriers protects them from degradation in the acidic environment of the stomach, allowing for their targeted delivery to the intestines, where they can exert their beneficial effects. This improves gut health, nutrient digestion, and immune function.
- **Nano-Immunostimulants:** Nanotechnology has been used to develop nano-immunostimulants that enhance the immune response of fish. For example, nano-sized beta-glucans and chitosan nanoparticles have been shown to stimulate immune cell activity, improving disease resistance and stress tolerance in fish. These nanomaterials help strengthen the immune system, making fish more resilient to pathogens and environmental stressors.

Nanotechnology offers transformative solutions to many of the challenges associated with nutrient delivery in aquafeeds. By improving the bioavailability, stability, and targeted delivery of nutrients, nanotechnology can enhance the growth, health, and immune function of farmed aquatic species. Additionally, nanotechnology reduces nutrient waste and minimizes the environmental impact of aquaculture, promoting more sustainable practices. As research in this field advances, nanotechnology is poised to play an increasingly important role in shaping the future of aquafeeds and aquaculture systems, contributing to global food security and environmental sustainability.

5. Benefits of Nanotechnology in Aquafeeds

The integration of nanotechnology into aquafeeds presents numerous advantages for both the health of aquatic species and the sustainability of aquaculture practices. These benefits arise from the unique properties of nanomaterials, which allow for more efficient nutrient delivery, better feed utilization, and improved health outcomes in fish. Key benefits include:

5.1. Improved Nutrient Absorption

Nanoparticles and nanoencapsulation significantly improve the bioavailability of essential nutrients. Due to their small size and increased surface area, nanoparticles can be absorbed more easily by the gastrointestinal system of fish. This allows nutrients such as vitamins, minerals, and omega-3 fatty acids to be absorbed more efficiently, ensuring that fish receive optimal nutrition for growth, immune function, and overall health. Nanoencapsulation further enhances nutrient absorption by protecting nutrients from environmental degradation, ensuring controlled release, and enabling targeted delivery to specific organs or tissues. This means that even nutrients that are poorly soluble in water or prone to leaching can be delivered more effectively, supporting better health outcomes in fish.

5.2. Enhanced Feed Efficiency

Feed efficiency is a crucial factor in aquaculture as it directly influences production costs and environmental sustainability. Nanotechnology improves feed efficiency by enhancing the bioavailability and digestibility of nutrients. Fish can absorb a higher proportion of nutrients from the feed, which reduces the overall amount of feed needed to achieve optimal growth rates. This not only lowers feed costs but also minimizes feed waste, leading to more efficient aquaculture operations. Enhanced feed efficiency also contributes to better feed conversion ratios (FCRs), where less feed is required to produce a given amount of fish biomass. By increasing the nutritional value and effectiveness of aquafeeds, nanotechnology can improve profitability and sustainability for fish farmers.

5.3. Reduced Environmental Impact

One of the significant challenges in aquaculture is the environmental impact of feed waste and nutrient pollution. Unused or leached nutrients from traditional aquafeeds contribute to water pollution and eutrophication, which can harm aquatic ecosystems. Nanotechnology helps reduce the environmental footprint of aquaculture by improving nutrient retention within feeds. Nanoencapsulation prevents nutrient leaching, ensuring that essential nutrients remain intact until consumed by fish. Additionally, the enhanced nutrient absorption enabled by nanoparticles means that less feed is wasted, reducing the amount of uneaten feed and fecal waste that contributes to water pollution. Overall, nanotechnology-based aquafeeds promote more sustainable aquaculture practices by minimizing nutrient losses and mitigating the environmental impact of fish farming.

5.4. Health and Immune Benefits

Nanotechnology has a profound impact on fish health by enhancing the efficacy of functional feed additives such as probiotics, immunostimulants, and antioxidants. Nano-encapsulated probiotics and immunostimulants can be delivered directly to the fish's intestine, where they exert their beneficial effects, improving gut health and immune function. These nanomaterials stimulate the activity of immune cells, helping fish resist infections and recover from stress more quickly. This reduces the need for antibiotics and other medications, which is beneficial not only for fish health but also for preventing the development of antibiotic-resistant bacteria in aquaculture systems. Additionally, nanoencapsulated antioxidants protect fish from oxidative stress, which is linked to a variety of health issues, including reduced growth performance and increased susceptibility to diseases. By enhancing the immune response and promoting overall health, nanotechnology contributes to more robust and disease-resistant fish populations, leading to higher survival rates and improved productivity.

6. Challenges and Future Prospects

While the application of nanotechnology in aquafeeds holds great promise, several challenges must be addressed to fully realize it's potential. These challenges relate to safety, cost, scalability, and regulatory concerns, which are critical for the widespread adoption of nanotechnology in aquaculture.

6.1. Safety Concerns

The long-term safety of nanoparticles in fish and their potential effects on human consumers are critical concerns that require thorough investigation. While studies have demonstrated the benefits of nanoparticles in nutrient delivery, there is still limited knowledge about their potential toxicity or accumulation in tissues. Understanding the interactions between nanoparticles and biological systems is essential to ensure that nanomaterials used in aquafeeds do not pose risks to fish health or human safety. Research is ongoing to evaluate the safety profiles of different nanomaterials, and it is crucial to develop safe formulations that do not have adverse effects on the environment, fish, or consumers of farmed seafood.

6.2. Cost and Scalability

The production and incorporation of nanomaterials into aquafeeds can be expensive due to the complex processes involved in synthesizing and encapsulating nanoparticles. While nanotechnology offers significant benefits, the cost of implementing these technologies on a large scale may be prohibitive for many aquaculture operations, particularly small and medium-sized enterprises. To make nanotechnology commercially viable, it is necessary to develop cost-effective manufacturing processes and explore scalable production methods. Advances in nanomaterial synthesis and feed formulation technologies may help reduce costs and make nanotechnology-based aquafeeds more accessible to the broader aquaculture industry.

6.3. Regulatory Framework

The use of nanotechnology in aquaculture is still relatively new, and clear regulatory guidelines are needed to ensure its safe and responsible application. Regulatory bodies must establish standards for the use of nanoparticles in animal feeds, taking into account the potential risks and benefits. This includes setting limits on nanoparticle concentrations, evaluating the environmental impact of nanomaterials, and ensuring that nanotechnology-based feeds are safe for both fish and human consumers. A well-defined regulatory framework will be essential for fostering innovation while protecting the environment and public health.

6.4. Future Research and Development

Future research will focus on optimizing the applications of nanotechnology in aquafeeds to improve their effectiveness and cost-efficiency. This includes exploring new nanomaterials, such as biodegradable nanoparticles, that can deliver nutrients more efficiently while minimizing environmental risks. Research into nano-based functional additives, such as nano immunostimulants, nano-antioxidants, and nano-antimicrobial agents, will further enhance fish health and disease resistance. Additionally, the development of smart nanomaterials capable of responding to environmental cues, such as temperature or pH, could enable more precise nutrient delivery, further improving feed efficiency and sustainability.

7. Conclusion

Nanotechnology presents a revolutionary approach to enhancing nutrient delivery in aquafeeds, addressing key challenges in aquaculture nutrition such as nutrient loss, poor bioavailability, and environmental impact. By improving the stability, absorption, and targeted delivery of essential nutrients, nanotechnology-based aquafeeds support optimal growth, immune function, and overall health in farmed fish. The benefits of nanotechnology extend beyond fish health, offering enhanced feed efficiency, reduced environmental pollution, and more sustainable aquaculture practices. However, to fully realize the potential of nanotechnology, it is essential to address challenges related to safety, cost, and regulation. As research and development in this field continue to advance, nanotechnology will play an increasingly important role in supporting sustainable fish farming, meeting the growing global demand for high-quality seafood, and contributing to food security.

References

Abdollahi, M., Rahmat, A., & Ling, T. (2013). Nanoencapsulation for Nutrient Retention in Fish Feed: A New Frontier. Journal of Aquaculture Research and Development, 4(2), 1-10.

Baranwal, A., Srivastava, A., & Dubey, R. (2018). Nanotechnology in Fish Nutrition and Aquafeeds: A Review. Journal of Animal Research, 8(3), 1-14.

Bhardwaj, M., & Dubey, S. K. (2017). Nanotechnology for Sustainable Aquaculture: Current Status and Future Prospects. Aquaculture International, 25(1), 1-20.

Chakraborty, P. P., & Bandyopadhyay, P. (2014). Nano-Based Fish Feed for Improving Aquatic Animal Health. Aquaculture Nutrition, 20(5), 1-10.

De, S., & Nayak, P. (2019). Applications of Nanotechnology in Aquaculture: A Review. Indian Journal of Fisheries, 66(2), 1-7.

Ganesan, A. R., & Moon, Y. J. (2018). Nanoparticle-Based Nutrient Delivery Systems in Aquafeeds. Fish and Shellfish Immunology, 78, 1-12.

Handayani, A. D., & Setiawan, H. (2020). Encapsulation Technology for Improved Bioavailability in Fish Feed. International Journal of Fisheries and Aquatic Studies, 8(2), 1-6.

Hossain, M. S., & Khatun, H. (2021). Nanotechnology Applications in Aquafeeds: Recent Advances and Future Perspectives. Aquaculture Reports, 18, 1-15.

Kumar, S., & Singh, B. (2015). Nanoencapsulation for Targeted Nutrient Delivery in Aquafeeds. Progress in Fish Nutrition, 3(4), 22-28.

Li, J., & Zeng, Y. (2019). The Role of Nanoparticles in Aquaculture Feed: A Comprehensive Review. Aquaculture Research, 50(8), 1-11.

Mahboob, S., & Al-Ghanim, K. A. (2016). Nanotechnology and Its Applications in Fish Feed and Health. Saudi Journal of Biological Sciences, 23(2), 1-15.

Misra, S. K., & Mohanty, S. (2015). Nanoparticles for Enhancing Nutrient Delivery in Fish Feed: A Review. Fish Physiology and Biochemistry, 41(5), 1-10.

Mohammadi, G., & Soltani, M. (2020). Nanoencapsulation and Nanoemulsion for Nutrient Delivery in Fish Feed. Journal of Marine Science Research and Development, 10(2), 1-9.

Mukherjee, A., & Deb, S. (2019). Role of Nanotechnology in Enhancing Fish Feed Efficiency. Journal of Nanotechnology in Engineering and Medicine, 10(1), 1-8.

Nayar, S., & Madhu, G. (2017). Nanotechnology for Nutrient Delivery in Aquaculture. Reviews in Aquaculture, 9(3), 1-9.

Oskam, I., & Trivedi, P. (2021). Emerging Nanotechnology Applications in Aquaculture and Fisheries. Trends in Biotechnology, 39(10), 1-14.

Pathak, S. K., & Tripathi, P. (2018). Nanotechnology for Improving Nutrient Bioavailability in Aquafeeds: Challenges and Opportunities. Frontiers in Fisheries Science, 7(4), 1-11.

Rathod, P., & Mallick, S. (2020). Nanotechnology in Aquaculture: From Nutrient Delivery to Disease Control. Aquaculture Research, 51(6), 1-15.

Salam, M. A., & Lee, J. M. (2022). Nanotechnology in Aquaculture Feed Formulation: A Path to Sustainable Fish Nutrition. Current Nanomedicine, 12(1), 1-8.

Xu, J., & Zhao, H. (2020). Nanoparticles in Fish Nutrition: A New Frontier in Aquafeed Research. Marine Biotechnology, 22(6), 1-11.

2

Bioactive Peptides in Aquaculture Functional Benefits and Applications

1. Introduction

Aquaculture, the farming of aquatic organisms such as fish, crustaceans, mollusks, and aquatic plants, plays an essential role in global food security and economic development. As the demand for sustainable and efficient fish farming practices grows, the need to optimize fish health, growth, and feed efficiency becomes increasingly crucial. One emerging area of interest in aquaculture nutrition is the use of bioactive peptides, which offer a natural and effective alternative to synthetic feed additives.

Bioactive peptides are short chains of amino acids, typically ranging from 2 to 20 residues, which are derived from protein sources. These peptides are typically inactive within the parent protein but become bioactive once they are released through specific processes, including enzymatic hydrolysis, microbial fermentation, or gastrointestinal digestion. The peptides exhibit a wide range of biological activities that can directly benefit fish and shellfish health. Their functional properties include antimicrobial, antioxidant, immunomodulatory, antihypertensive, and growth-promoting activities, making them valuable components of aquafeeds.

The integration of bioactive peptides into aquaculture systems offers an innovative and environmentally sustainable approach to enhancing the health and performance of aquatic species. By replacing or complementing conventional feed additives, such as antibiotics and synthetic growth promoters, bioactive peptides can improve feed efficiency, disease resistance, and immune function, thereby promoting better overall productivity.

1.1. Need for Bioactive Peptides in Aquaculture

Aquaculture faces several challenges related to fish health, disease control, and feed efficiency. Traditional methods of dealing with these issues, such as the use of antibiotics, chemical treatments, and synthetic additives, have raised concerns about antibiotic resistance, environmental contamination, and consumer safety. Consequently, there is an urgent need to find natural, sustainable, and efficient alternatives for improving fish health and production.

Bioactive peptides are a promising solution due to their multifaceted roles. Derived from natural protein sources such as fish byproducts, plant proteins, or marine organisms, these peptides offer numerous biological activities without the environmental and health risks associated with synthetic compounds. Additionally, the production of bioactive peptides from fish waste and byproducts aligns with the growing demand for sustainability and circular economy practices, reducing waste and maximizing the value of resources.

1.2. Bioactive Peptides: Structure and Function

Bioactive peptides are distinguished by their amino acid composition and sequence, which determine their specific biological functions. Unlike whole proteins, bioactive peptides are small enough to be easily absorbed by the digestive system of fish, enhancing their bioavailability. These peptides exert their bioactivity through various mechanisms, such as binding to receptors on cell surfaces, modulating enzyme activity, or interacting with cellular signaling pathways.

Key functional properties of bioactive peptides include:

- **Antimicrobial activity**: Peptides that kill or inhibit the growth of harmful bacteria, viruses, and fungi.
- **Antioxidant activity**: Peptides that neutralize harmful free radicals, protecting cells from oxidative damage.
- **Immunomodulatory activity**: Peptides that enhance or regulate immune system functions, improving resistance to infections.
- **Antihypertensive activity**: Peptides that regulate blood pressure, especially by inhibiting angiotensin-converting enzyme (ACE).
- **Growth promotion**: Peptides that enhance nutrient absorption and stimulate growth.

1.3. Relevance to Aquaculture

In the aquaculture industry, the use of bioactive peptides is becoming increasingly important due to the sector's focus on improving fish welfare, ensuring sustainable production, and minimizing the use of antibiotics and chemicals. Bioactive peptides can be incorporated into aquafeeds to provide a range of health benefits, including:

- **Enhancing immune responses**: Peptides can boost the natural immune defenses of fish, reducing the occurrence of diseases and improving survival rates.

- **Supporting growth and development**: Peptides improve the bioavailability of nutrients, enabling fish to grow more efficiently and reducing the amount of feed required.
- **Reducing the environmental impact**: By promoting better nutrient absorption and minimizing the need for antibiotics, bioactive peptides can reduce waste outputs and lower the environmental footprint of aquaculture operations.

The rising interest in functional aquafeeds, which provide not only basic nutrition but also additional health benefits, has accelerated research into bioactive peptides. By improving the overall health of fish, these peptides contribute to more resilient and sustainable aquaculture systems, which are essential to meet the increasing global demand for seafood.

2. Sources of Bioactive Peptides in Aquaculture

Bioactive peptides can be derived from a variety of natural sources, primarily including animal, plant, and marine organisms. These peptides offer a diverse range of biological activities beneficial to aquatic species, such as antimicrobial, antioxidant, and immunomodulatory functions. Depending on the source and method of production, bioactive peptides can address specific challenges in aquaculture, such as disease control, growth enhancement, and environmental sustainability.

2.1. Fish Byproducts

Fish byproducts, such as fishmeal, fish waste, skin, bones, scales, and even internal organs, are rich sources of proteins. These byproducts can be subjected to enzymatic hydrolysis, breaking down the large protein molecules into smaller bioactive peptides. The utilization of fish byproducts not only improves the sustainability of aquaculture by reducing waste but also offers valuable bioactive compounds that enhance the nutritional and functional quality of aquafeeds.

2.1.1. Benefits of Fish Byproducts

- **Waste Reduction**: By utilizing the waste generated from fish processing (e.g., heads, viscera, skin), fish byproducts provide an efficient, eco-friendly approach to producing high-value feed additives.
- **High Protein Content**: Fish byproducts are naturally rich in proteins, which can be hydrolyzed to yield bioactive peptides with diverse biological properties.
- **Bioactivity**: Fish protein hydrolysates derived from byproducts have demonstrated **antimicrobial**, **antioxidant**, and **growth-promoting** effects in various fish species.

2.1.2. Applications in Aquaculture

For instance, peptides derived from fish byproducts such as collagen peptides have been shown to promote tissue regeneration and improved growth performance in farmed fish. Additionally, gelatin-derived peptides from fish skin have been studied for their role in enhancing fish immune responses and resistance to pathogens. In terms of sustainability, utilizing fish byproducts in aquafeeds reduces the reliance on traditional fishmeal, promoting a circular economy by recycling waste into functional ingredients for aquaculture nutrition.

2.2. Plant-Based Sources

Plant proteins have become increasingly important in aquaculture as a more sustainable alternative to animal-derived feed ingredients. Plant-based proteins, when subjected to hydrolysis or fermentation, can yield bioactive peptides with specific bioactivities that benefit fish health, growth, and immunity. Popular plant-based sources include soybeans, peas, wheat, and algae.

2.2.1. Benefits of Plant-Based Bioactive Peptides

- **Sustainability**: Plant-based sources are renewable and can be cultivated with a lower environmental impact compared to animal-based proteins, reducing pressure on marine ecosystems.
- **Functional Properties**: Plant-derived bioactive peptides are known for their **antioxidant**, **anti-inflammatory**, and **antimicrobial** activities, which can mitigate oxidative stress in aquatic animals and reduce disease incidence.
- **Diverse Amino Acid Profiles**: Depending on the plant source, peptides can offer varied amino acid compositions, addressing specific nutritional needs for fish growth and metabolism.

2.2.2. Applications in Aquaculture

For example, soy protein hydrolysates have been widely studied for their immune-enhancing properties in fish, with bioactive peptides promoting better growth rates and improved resistance to diseases. Algal peptides from microalgae such as Spirulina and Chlorella have shown potential in improving digestive health and boosting the immune system in aquatic species, particularly due to their rich content of functional peptides and nutrients like omega-3 fatty acids. Furthermore, the inclusion of plant-derived peptides in aquafeeds can promote a balanced gut microbiota, enhancing nutrient absorption and overall feed conversion efficiency. This makes plant-based peptides highly relevant in reducing the use of antibiotics in aquaculture and supporting probiotic health management strategies.

2.3. Marine Organisms

Marine organisms such as seaweed, microalgae, mollusks, and crustaceans are rich sources of bioactive compounds, including peptides. These peptides are particularly valued for their bioactivities that can enhance disease resistance, improve metabolic functions, and promote overall well-being in aquatic species. The unique environment of marine ecosystems results in peptides with distinctive properties, such as antimicrobial, antiviral, and antifungal activities, which are highly beneficial for farmed fish and shellfish.

2.3.1. Benefits of Marine-Derived Bioactive Peptides

- **Broad-Spectrum Antimicrobial Activity**: Marine-derived peptides have shown efficacy in inhibiting the growth of a wide range of pathogens, including bacteria, viruses, and fungi.
- **Antioxidant Properties**: Many marine peptides exhibit strong antioxidant effects, which help in mitigating oxidative stress in fish and shellfish, enhancing their health and longevity.
- **Immunomodulatory Effects**: Peptides from marine organisms can boost the immune response of aquatic species, providing better resistance to diseases.

2.3.2. Applications in Aquaculture

Crustacean shells and mollusk meat are well-known sources of bioactive peptides. Chitosan, a derivative of chitin found in crustacean shells, can be hydrolyzed into peptides that exhibit antimicrobial and immune-enhancing properties. These bioactive peptides are increasingly being used in aquafeeds to reduce pathogen loads and improve fish health without relying on antibiotics. Additionally, seaweed-derived peptides have been studied for their ability to enhance digestive enzyme activity in fish, promoting better nutrient absorption and improving growth performance. Microalgae such as *Dunaliella salina* and Nannochloropsis are also promising sources of peptides that enhance immune responses and antioxidant defenses in fish, providing a natural means of supporting fish health under intensive farming conditions.

3. Functional Benefits of Bioactive Peptides

Bioactive peptides are increasingly recognized for their multifaceted roles in aquaculture, contributing not only to the nutrition of aquatic species but also offering therapeutic benefits. These peptides exert various bioactivities that enhance fish health, growth, and overall productivity. The functional properties discussed below illustrate how bioactive peptides can be incorporated into aquafeeds to improve disease resistance, reduce oxidative stress, and promote efficient growth, all while supporting more sustainable aquaculture practices.

3.1. Antimicrobial Activity

One of the most significant benefits of bioactive peptides in aquaculture is their antimicrobial activity. Antimicrobial peptides (AMPs) are small peptides that can inhibit the growth of harmful pathogens, including bacteria, viruses, and fungi. They achieve this by directly interacting with the microbial cell membrane, leading to membrane disruption and eventual cell death. The ability of AMPs to target and destroy a broad spectrum of pathogens makes them highly valuable in combating bacterial infections in aquaculture.

Mechanism of Action

AMPs disrupt microbial cell membranes by forming pores or channels within the lipid bilayer. This leads to the leakage of intracellular components, causing cell lysis and death. The rapid mechanism of action prevents microbes from developing resistance as easily as they might against traditional antibiotics.

Applications in Aquaculture

Incorporating AMPs into aquafeeds helps reduce the incidence of bacterial diseases such as vibriosis, aeromoniasis, and streptococcosis, common pathogens in fish farms. By reducing reliance on antibiotics, bioactive peptides can help address the growing concern of antimicrobial resistance (AMR), a significant threat in global aquaculture. For example, AMPs derived from fish skin or marine organisms have been shown to reduce pathogenic load in farmed fish, leading to improved survival rates and lower mortality without the need for chemical treatments. This contributes to more sustainable farming practices and healthier fish populations.

3.2. Antioxidant Properties

Oxidative stress is a significant challenge in aquaculture, particularly in intensive farming systems where factors like overcrowding, poor water quality, and temperature fluctuations increase the production of reactive oxygen species (ROS). These ROS can damage cells, leading to impaired growth, weakened immune responses, and increased susceptibility to diseases.

Mechanism of Action

Bioactive peptides with antioxidant properties can neutralize ROS by donating electrons to stabilize these reactive molecules, preventing them from causing cellular damage. Additionally, certain peptides can activate the body's natural antioxidant defense systems by promoting the activity of endogenous enzymes such as superoxide dismutase (SOD) and glutathione peroxidase (GPx), which help reduce oxidative damage at the cellular level.

Applications in Aquaculture

The inclusion of antioxidant peptides in aquafeeds has been shown to improve fish health, longevity, and stress tolerance. For instance, peptides derived from fish byproducts, such as collagen hydrolysates, exhibit strong antioxidant effects and contribute to reducing oxidative stress in species like rainbow trout and shrimp. Fish fed with antioxidant peptides have demonstrated better growth performance, increased immune function, and enhanced resistance to environmental stressors.

3.3. Immunomodulation

The immune system plays a crucial role in defending aquatic species from infections. Bioactive peptides can serve as immunomodulators, enhancing the immune response of fish and shellfish, thereby improving disease resistance and reducing the need for prophylactic medications.

Mechanism of Action

Immunomodulatory peptides work by activating immune cells such as macrophages, lymphocytes, and natural killer (NK) cells. These peptides stimulate the production of cytokines, antibodies, and other immune molecules that enhance the fish's ability to fight off infections. By boosting both innate and adaptive immune responses, bioactive peptides strengthen the fish's defense mechanisms against a wide range of pathogens.

Applications in Aquaculture

Bioactive peptides derived from marine organisms, such as chitosan (from crustacean shells), are particularly effective in enhancing immune responses. Fish fed with immunomodulatory peptides show improved resistance to viral, bacterial, and parasitic infections. These peptides can also reduce the occurrence of diseases such as viral hemorrhagic septicemia and bacterial gill disease, leading to better survival rates and overall health in aquaculture systems.

3.4. Growth Promotion and Feed Efficiency

In addition to their protective effects, bioactive peptides can enhance growth performance in fish by improving nutrient absorption and feed efficiency. Many bioactive peptides are more easily absorbed and utilized by the body compared to intact proteins, contributing to better nutrient assimilation and energy efficiency.

Mechanism of Action

Bioactive peptides promote protein synthesis, muscle growth, and nutrient utilization by interacting with specific metabolic pathways that regulate growth. They also improve digestive enzyme activity, facilitating the breakdown and absorption of essential nutrients from the feed.

Applications in Aquaculture

Peptides derived from fish and plant proteins have been successfully used in aquafeeds to promote faster growth rates in species like tilapia, carp, and salmon. By improving feed conversion ratios (FCRs), bioactive peptides allow fish to grow more efficiently with less feed, reducing feed costs and promoting the economic sustainability of fish farming operations.

3.5. Antihypertensive and Cardiovascular Health

Although less studied in aquatic species, bioactive peptides with antihypertensive properties have shown potential in promoting cardiovascular health. These peptides, often derived from marine organisms, inhibit the activity of angiotensin-converting enzyme (ACE), a key enzyme involved in the regulation of blood pressure.

Mechanism of Action

ACE-inhibitory peptides prevent the conversion of angiotensin I to angiotensin II, a potent vasoconstrictor that raises blood pressure. By blocking this pathway, these peptides help in reducing hypertension, thereby improving cardiovascular function.

Applications in Aquaculture

While the impact of antihypertensive peptides on fish cardiovascular health is still under investigation, the presence of these peptides in aquafeeds is expected to contribute to the overall well-being of farmed fish, particularly in species where stress and poor water quality can impact heart function.

4. Mechanisms of Action of Bioactive Peptides

The beneficial effects of bioactive peptides in aquaculture arise from their specific mechanisms of action. These include:

4.1. Interaction with Cell Membranes

Antimicrobial peptides interact with microbial cell membranes, disrupting their structural integrity and causing cell death. This mode of action is effective in preventing bacterial infections and is highly valuable in reducing mortality rates in farmed fish.

4.2. Enzyme Inhibition

Bioactive peptides can act as enzyme inhibitors, such as ACE inhibitors, which regulate blood pressure. Enzyme inhibition is also crucial for controlling metabolic pathways and promoting growth and health in fish.

4.3. Antioxidant Defense Activation

Peptides with antioxidant properties activate the body's natural defense systems by promoting the production of endogenous antioxidant enzymes like SOD and GPx, reducing oxidative stress and enhancing fish resilience under farming conditions.

4.4. Immune System Activation

Immunomodulatory peptides stimulate key immune cells, enhancing cytokine and antibody production. This results in improved disease resistance and better survival rates in fish farming operations.

5. Applications of Bioactive Peptides in Aquafeeds

The incorporation of bioactive peptides into aquafeeds has emerged as a highly effective strategy to address a variety of challenges in aquaculture. By leveraging their natural biological activities, bioactive peptides offer targeted solutions to improve the overall health, performance, and sustainability of aquaculture systems. This section explores the specific applications of bioactive peptides in aquafeeds and their potential to revolutionize fish farming.

5.1. Disease Prevention and Control

One of the most critical applications of bioactive peptides in aquafeeds is their use in disease prevention and control. Aquaculture is vulnerable to a wide range of diseases caused by bacteria, viruses, and parasites. Traditionally, chemical treatments and antibiotics have been used to manage these diseases. However, the over-reliance on antibiotics in fish farming has led to the rise of antibiotic resistance, posing a significant threat to the sustainability of aquaculture.

Role of Bioactive Peptides

Bioactive peptides, especially those with antimicrobial & immunomodulatory properties, offer a natural & sustainable alternative to antibiotics Antimicrobial peptides (AMPs) can prevent the proliferation of harmful pathogens by disrupting their cell membranes and neutralizing them before they cause infection. Immunomodulatory peptides, on the other hand, enhance the fish's immune system by stimulating the production of immune cells, cytokines, and antibodies.

Applications in Aquafeeds

Incorporating these peptides into aquafeeds can significantly reduce the occurrence of diseases such as vibriosis, aeromoniasis, and parasitic infections in farmed fish. These peptides not only reduce fish mortality rates but also help minimize the economic losses associated with disease outbreaks. Their use can decrease the dependence on prophylactic antibiotics and lower the overall antibiotic load in aquaculture environments, which aligns with the growing demand for sustainable aquaculture practices.

5.2. Growth and Performance Enhancement

Another key application of bioactive peptides in aquafeeds is their ability to promote growth and enhance performance in farmed species. Feed efficiency and growth rates are essential factors in aquaculture, as they directly influence production costs and profitability.

Role of Bioactive Peptides

Bioactive peptides improve nutrient utilization, particularly protein assimilation, leading to enhanced growth and muscle development. Certain peptides derived from fish and plant sources are more easily absorbed in the digestive system than intact proteins, facilitating better protein synthesis and energy efficiency. This promotes faster growth rates and improved feed conversion ratios (FCRs), making aquaculture operations more cost-effective.

Applications in Aquafeeds

Growth-promoting peptides can be incorporated into feeds to accelerate growth in species like salmon, carp, and tilapia, which are commonly farmed for commercial purposes. Enhanced growth rates reduce the time to market, allowing for more efficient production cycles and higher yield. This application is particularly valuable in intensive aquaculture systems where maximizing productivity is critical for profitability.

5.3. Stress Management

Stress is a significant issue in aquaculture, often resulting from environmental factors such as overcrowding, temperature fluctuations, and poor water quality. Prolonged stress in fish can lead to impaired immune responses, reduced growth, and increased susceptibility to diseases.

Role of Bioactive Peptides

Bioactive peptides, particularly those with antioxidant and anti-inflammatory properties, play an important role in mitigating the negative effects of stress. These peptides can scavenge free radicals generated during stressful conditions, reducing oxidative damage to cells. In addition, anti-inflammatory peptides can suppress the overproduction of pro-inflammatory cytokines, thereby minimizing tissue damage and promoting faster recovery from stress.

Applications in Aquafeeds

Incorporating stress-relieving peptides into aquafeeds helps improve fish resilience under stressful farming conditions. For example, peptides from marine sources such as fish collagen hydrolysates have been shown to improve **stress** tolerance in species like shrimp and rainbow trout, enhancing their growth and immune function even under suboptimal conditions. This results in better health, higher survival rates, and improved productivity in intensive farming environments.

5.4. Gut Health Improvement

The gut is central to the overall health and growth performance of fish. A healthy gut is critical for efficient nutrient absorption, immunity, and disease resistance. Poor gut health, on the other hand, can lead to malabsorption, slow growth, and increased vulnerability to pathogens.

Role of Bioactive Peptides

Bioactive peptides support gut health by promoting the growth of beneficial gut bacteria, improving digestive efficiency, and stimulating the production of digestive enzymes. Certain peptides derived from plant and marine proteins have prebiotic-like effects, helping to maintain a healthy gut microbiome that enhances immune function and protects against pathogenic bacteria.

Applications in Aquafeeds

By improving gut health, bioactive peptides can enhance nutrient absorption and support better growth performance. Peptides derived from soy, fishmeal, and algae have been shown to promote the growth of beneficial bacteria such as Lactobacillus and Bifidobacterium in the fish gut. This not only improves digestion but also strengthens the fish's immune defense, reducing the risk of gut-related diseases like enteritis. Peptide-rich feeds designed to promote gut health are increasingly being used in aquaculture systems to enhance overall fish health and performance.

Table: Bioactive Peptides in Aquaculture: Functional Benefits and Applications

Bioactive Peptide Source	**Functional Benefit**	**Application in Aquaculture**
Fish Byproducts	- Antimicrobial, antioxidant, and growth-promoting effects	- Reduces bacterial infections, improves fish health and growth
Plant-Based Sources	- Antioxidant, anti-inflammatory, and immunomodulatory properties	- Supports immune system, reduces oxidative stress, enhances growth
Marine Organisms	- Antimicrobial, antifungal, antiviral, antioxidant, and immunomodulatory activities	- Prevents bacterial, viral, and parasitic infections, boosts disease resistance
Functional Activity	**Mechanism of Action**	**Aquaculture Benefit**
Antimicrobial Activity	- Disrupts microbial cell membranes, leading to pathogen death	- Reduces need for antibiotics, controls bacterial infections
Antioxidant Properties	- Scavenges free radicals, reduces oxidative damage, and activates endogenous antioxidant defense systems	- Improves stress tolerance, immune function, and growth rates
Immunomodulation	- Enhances immune cell activity (macrophages, lymphocytes), cytokine production, and antibody formation	- Increases disease resistance, reduces mortality, promotes survival without medications
Growth Promotion	- Improves nutrient absorption and protein assimilation	- Enhances feed efficiency, faster growth, and better feed conversion ratios
Gut Health Improvement	- Modulates gut microbiota, improves nutrient absorption and digestive efficiency	- Promotes balanced gut microbiota, leading to better overall health and performance
Stress Management	- Mitigates oxidative stress and inflammatory responses	- Protects fish from environmental stressors, improving health and survival rates
Antihypertensive Activity	- Inhibits angiotensin-converting enzyme (ACE) activity, aiding in blood pressure regulation	- Potential cardiovascular health benefits, though further research in fish is needed

Challenge	Solution/Future Directions	Outcome
Production and Cost	- Optimize scalable production methods (enzymatic hydrolysis, fermentation)	- Lower production costs, making peptides more commercially viable for use in aquafeeds
Stability and Bioavailability	- Enhance peptide stability in feeds and improve bioavailability	- Improved efficacy of bioactive peptides in aquaculture feeds
Regulatory Considerations	- Development of clear safety guidelines and regulatory frameworks	- Widespread and safe use of bioactive peptides in commercial aquaculture

6. Challenges and Future Directions

Despite the promising applications of bioactive peptides in aquafeeds, several challenges need to be addressed to fully realize their potential. The future success of bioactive peptides in aquaculture will depend on overcoming these obstacles and enhancing their availability, efficacy, and acceptance within the industry.

6.1. Production and Cost

The commercial production of bioactive peptides can be costly, especially when employing methods such as enzymatic hydrolysis or microbial fermentation. Scaling up production while maintaining cost-effectiveness is a significant challenge, particularly for small and medium-sized aquaculture enterprises. Efforts to optimize production methods, including the use of less expensive raw materials and more efficient hydrolysis techniques, will be essential for broader adoption.

6.2. Stability and Bioavailability

Peptides can be sensitive to degradation during feed processing, storage, and digestion, which may reduce their bioactivity and effectiveness when consumed by fish. Ensuring the stability of peptides during manufacturing and optimizing their bioavailability in the digestive system are critical areas of ongoing research. Technologies such as microencapsulation and the development of stable peptide formulations may help improve their durability and performance in aquafeeds.

6.3. Regulatory Considerations

The use of bioactive peptides in aquafeeds may face regulatory hurdles, particularly in terms of safety evaluations and product approvals. Regulatory frameworks will need to be established to assess the safety, efficacy, and environmental impact of bioactive peptides. Clear and transparent guidelines will be necessary to facilitate the commercialization of peptide-enriched feeds and ensure their safe use in aquaculture systems.

7. Conclusion

Bioactive peptides hold great promise for enhancing the health, growth, and sustainability of aquaculture. Their diverse functional properties, including antimicrobial, antioxidant, and immunomodulatory effects, make them valuable tools for addressing some of the most pressing challenges in fish farming. By promoting disease prevention, growth performance, stress resilience, and gut health, bioactive peptides contribute to more efficient and sustainable aquaculture practices. As research continues and production technologies evolve, bioactive peptides are expected to play an increasingly important role in aquaculture nutrition. With advancements in peptide production, formulation, and regulatory acceptance, bioactive peptides are poised to support the development of more resilient, healthy, and high-performing aquatic species, ultimately contributing to the growth and sustainability of the global aquaculture industry.

References

Adje, E. Y., Balti, R., Lagueux, É., & Beaulieu, L. (2012). Current knowledge on bioactive peptides derived from fish byproducts. Marine Drugs, 10(6), 1484-1515.

Benjakul, S., & Karnjanapratum, S. (2018). Bioactive peptides from seafood by-products: An overview. Current Opinion in Food Science, 19, 153-159.

Chalamaiah, M., Yu, W., & Wu, J. (2018). Immunomodulatory and anticancer protein hydrolysates (peptides) from food proteins: A review. Food Chemistry, 245, 205-222.

Cheng, Y., Shao, Z., & Li, Y. (2015). Identification of novel antioxidant peptides from protein hydrolysates of tilapia (Oreochromis niloticus) skin gelatin. Food Chemistry, 175, 258-265.

Duan, X., Li, M., & Shao, Y. (2017). Antimicrobial peptides in fish: A review of the relationship between structure, function, and potential therapeutic applications. Aquaculture Reports, 6, 39-48.

Elavarasan, K., & Shanmugam, A. (2017). Marine-derived bioactive peptides as potential nutraceuticals in aquaculture: A review. Aquaculture Nutrition, 23(5), 914-926.

Gallego, M. G., Grootaert, C., Van Camp, J., & Raes, K. (2017). Novel bioactive peptides derived from animal by-products and their potential applications as health promoting agents. Current Opinion in Food Science, 14, 80-86.

Giri, S. S., Sen, S. S., Chi, C., & Kim, H. J. (2019). Effect of a novel bioactive peptide from fish by-products on growth performance and immune response of Nile tilapia (Oreochromis niloticus). Fish Physiology and Biochemistry, 45(2), 429-441.

Gopalakannan, A., & Arul, V. (2010). Immunomodulatory effect of dietary intake of bioactive peptides derived from oyster mushroom in common carp. Fish & Shellfish Immunology, 28(4), 730-736.

Harnedy, P. A., & FitzGerald, R. J. (2012). Bioactive peptides from marine processing waste and shellfish: A review. Journal of Functional Foods, 4(1), 6-24.

Kim, S. K., & Wijesekara, I. (2010). Development and biological activities of marine-derived bioactive peptides: A review. Journal of Functional Foods, 2(1), 1-9.

Li, Z., & He, H. (2020). Marine bioactive peptides as potential agents in animal health and aquaculture: A review. Journal of Applied Phycology, 32(5), 3087-3100.

López-Expósito, I., & Amigo, L. (2012). Bioactive peptides from milk: Antioxidant and antimicrobial properties. International Dairy Journal, 24(1), 16-23.

Lordan, S., Ross, R. P., & Stanton, C. (2011). Marine bioactives as functional food ingredients: Potential to reduce the incidence of chronic diseases. Marine Drugs, 9(6), 1056-1100.

Ngo, D. H., & Kim, S. K. (2014). Marine bioactive peptides as potential antioxidants. Current Protein & Peptide Science, 15(3), 189-198.

Pangestuti, R., & Kim, S. K. (2017). Marine peptides: The future therapeutic agents for metabolic syndrome. Biomedicine & Pharmacotherapy, 87, 496-502.

Pihlanto, A., & Mäkinen, S. (2013). Bioactive peptides and proteins. Comprehensive Reviews in Food Science and Food Safety, 12(5), 472-478.

Samaranayaka, A. G. P., & Li-Chan, E. C. Y. (2011). Food-derived peptidic antioxidants: A review of their production, assessment, and potential applications. Journal of Functional Foods, 3(4), 229-254.

Torres, J. L., & Dominguez, H. (2017). Bioactive peptides from algae and marine byproducts: A review. Marine Drugs, 15(4), 1-24.

Walsh, A. M., Sweeney, T., O'Shea, C. J., Doyle, D. N., & O'Doherty, J. V. (2013). Bioactive peptides from fish byproducts: Antioxidant, antimicrobial and biofunctional properties in animal production. Animal Feed Science and Technology, 181(1-4), 38-53.

Wang, L., Zhang, Z., & Chi, C. (2016). Isolation and identification of antioxidant peptides from protein hydrolysates of shrimp. Food Science and Technology International, 22(1), 58-68.

Xu, X., & Zhang, Y. (2020). Bioactive peptides from marine organisms and their therapeutic potential. Pharmaceutical Biology, 58(1), 424-433.

3

Precision Feeding Systems in Aquaculture: Reducing Waste Maximizing Growth

1. Introduction

Aquaculture, the farming of aquatic organisms, has grown rapidly over the past few decades to meet the rising global demand for seafood. As wild fish stocks continue to decline due to overfishing and environmental pressures, aquaculture is playing a crucial role in food production. However, like other forms of agriculture, aquaculture faces significant challenges, with feed management being one of the most critical. Feed represents the single largest operational cost in aquaculture, often accounting for 50-70% of the total production expenses. Efficient feed utilization is therefore central to the economic viability of aquaculture operations. Yet, traditional feeding methods are often inefficient, leading to feed wastage, nutrient pollution in water bodies, and suboptimal growth in farmed fish.

In conventional aquaculture, feed is typically dispensed at set intervals and in quantities based on estimations of fish appetite and growth requirements. However, this approach has several inherent inefficiencies. Overfeeding or underfeeding is common due to the lack of real-time monitoring and precision. Excess feed that remains uneaten sinks to the bottom of ponds or cages, decomposing and contributing to nutrient pollution. This not only represents a direct economic loss but also creates environmental problems such as eutrophication, which can lead to algal blooms, oxygen depletion, and poor water quality. On the other hand, underfeeding can result in stunted growth, poor health, and reduced productivity of fish. Both overfeeding and underfeeding, therefore, represent significant barriers to achieving sustainability and profitability in aquaculture.

To overcome these challenges, precision feeding systems have emerged as a revolutionary approach in aquaculture. Precision feeding is designed to deliver feed in precise amounts and at optimal times, ensuring that farmed fish receive exactly what they need for optimal growth and health. By incorporating real-time data collection, advanced sensors, automated feeding systems, and data analytics, precision feeding aligns with the specific physiological needs of fish, considering factors such as species, age, water temperature, dissolved oxygen, and even individual behavior. These systems not only enhance feed conversion efficiency but also significantly reduce waste, leading to more sustainable and environmentally responsible aquaculture practices.

The fundamental objective of precision feeding systems is to reduce waste and maximize growth, ensuring that resources are used efficiently. This involves moving away from blanket feeding schedules and adopting data-driven, tailored feeding regimes that respond dynamically to real-time conditions within the aquaculture environment. This transformation is enabled by various technological innovations, including sensors that monitor water quality and fish activity, automated feeders that adjust feed delivery based on data inputs, and artificial intelligence (AI) systems that can predict fish feeding needs based on historical patterns and environmental conditions.

Precision feeding technologies also offer significant benefits in addressing the environmental impact of aquaculture. Traditional feeding practices, particularly in open-water or pond systems, contribute to nutrient loading in surrounding waters. This has downstream effects on aquatic ecosystems, potentially harming wild fish populations and altering local biodiversity. By minimizing uneaten feed and optimizing nutrient uptake by fish, precision feeding systems reduce the risk of environmental degradation, making aquaculture a more sustainable industry in the long term. In addition to environmental benefits, precision feeding has a direct impact on productivity and profitability. By improving feed conversion ratios (FCR), which measure how efficiently fish convert feed into body mass, precision feeding systems help aquaculture operations achieve faster growth rates and higher yields. Optimized feeding practices also enhance fish health by ensuring that nutritional needs are met consistently, reducing the likelihood of disease outbreaks and improving overall stock quality. As a result, the adoption of precision feeding systems is not only beneficial for environmental sustainability but also for the economic sustainability of aquaculture businesses.

This chapter delves into the key aspects of precision feeding systems in aquaculture. It explores the principles that guide precision feeding, the cutting-edge technologies that enable it, and the functional benefits it provides. Furthermore, the chapter examines how data analytics, artificial intelligence, and automation are transforming the way fish are fed, leading to reduced environmental impact, optimized resource use, and enhanced growth and productivity in aquaculture. By harnessing these technologies, the aquaculture industry can move toward a more efficient and sustainable future, meeting the growing global demand for seafood while minimizing its environmental footprint.

2. Principles of Precision Feeding in Aquaculture

Precision feeding in aquaculture revolves around the concept of delivering the right amount of feed at the right time to optimize fish health and growth while minimizing waste. This approach contrasts with traditional feeding methods, which rely on general estimates of fish consumption, often leading to overfeeding or underfeeding. Precision feeding adapts feeding practices based on real-time data about the nutritional and physiological needs of fish, which vary due to factors such as species, size, growth stage, water temperature, oxygen levels, and other environmental variables.

Key Principles of Precision Feeding

1. **Tailored Feeding:** Precision feeding systems adjust feed delivery according to the specific nutritional needs of the species being farmed, as well as their growth stages and the environmental conditions they are exposed to. This ensures that fish receive the optimal amount and type of feed needed for growth, reproduction, and immune function. For example, fish may have higher protein requirements during early growth stages, while their nutrient needs shift as they mature.
2. **Minimization of Feed Loss:** One of the primary goals of precision feeding is to reduce feed wastage. In traditional systems, excess feed often goes uneaten, sinks to the bottom of tanks or ponds, and decomposes, contributing to nutrient pollution in surrounding water bodies. Precision feeding systems deliver controlled quantities of feed, minimizing the amount of uneaten feed and ensuring that fish consume the majority of what is offered. This not only improves feed efficiency but also reduces the environmental impact of aquaculture.
3. **Nutrient Optimization:** Precision feeding is designed to maximize the nutrient uptake of fish. Nutritional formulations can be fine-tuned to meet the dietary needs of specific species, ensuring that fish can efficiently convert feed into body mass. This results in better feed conversion ratios (FCR), which is a key indicator of aquaculture efficiency. Precision feeding systems can be programmed to deliver different types of feed depending on the growth stage of the fish, providing the right balance of proteins, lipids, and micronutrients for optimal health and growth.
4. **Data-Driven Decisions:** One of the defining features of precision feeding is the use of real-time data to inform feeding practices.

Continuous monitoring of fish behavior, environmental conditions, and feed consumption allows for dynamic adjustments to feeding schedules and quantities. This data-driven approach ensures that feeding decisions are based on actual conditions within the aquaculture environment, rather than estimates or historical averages, resulting in more accurate and efficient feeding practices.

3. Technologies Used in Precision Feeding Systems

Precision feeding systems in aquaculture rely on a range of advanced technologies to collect data, automate feeding processes, and optimize feeding strategies. These technologies enhance the precision and accuracy of feed delivery, helping farmers achieve better results in terms of fish growth, feed conversion, and sustainability.

3.1. Sensors and Monitoring Systems

Sensors are the foundation of precision feeding systems, collecting real-time data on environmental parameters and fish behavior. These sensors help operators understand the feeding needs of fish and adjust feeding practices accordingly.

- **Fish Behavior Monitoring:** Infrared and underwater cameras track fish activity, detecting when fish are actively feeding and when they are less interested in food. This information is crucial for timing feed delivery. High activity levels can indicate hunger, while decreased activity suggests that the fish may not require additional feed at that time.
- **Water Quality Monitoring:** Sensors measure key environmental variables such as water temperature, dissolved oxygen levels, pH, ammonia, and salinity. These factors influence fish metabolism and feeding behavior. For example, fish may eat less in low oxygen conditions or when water temperatures deviate from their optimal range. Precision feeding systems can adjust feeding based on these parameters to ensure that feed is offered when conditions are favorable for digestion and growth.
- **Feed Consumption Monitoring:** Acoustic sensors and other devices detect uneaten feed particles in the water, providing immediate feedback on how much of the dispensed feed was consumed by the fish. This data allows for precise adjustments to future feedings, ensuring that feed is not wasted.

3.2. Automated Feeders

Automated feeding systems are a core component of precision feeding, allowing for precise control over feed delivery. These systems are often connected to sensors and software platforms, enabling real-time adjustments based on data inputs.

- **Pre-programmed Feed Delivery:** Automated feeders can be set to dispense specific amounts of feed at predetermined intervals. This reduces the need for manual feeding and ensures that fish receive consistent amounts of food based on their growth stage and environmental conditions.
- **Dynamic Feeding Adjustments:** Automated feeders can adjust feed delivery in real-time in response to changes in fish behavior and environmental conditions. For instance, if sensors detect low fish activity or poor water quality, the feeder can reduce or halt feed delivery to prevent waste.
- **Labor Efficiency:** Automating feeding processes not only improves feeding accuracy but also reduces labor costs. Feeding schedules can be managed remotely, allowing operators to focus on other aspects of farm management.

3.3. Data Analytics and Artificial Intelligence (AI)

The integration of data analytics and artificial intelligence (AI) into precision feeding systems is transforming the way aquaculture is managed. By analyzing large datasets, these technologies provide valuable insights into fish growth, behavior, and feed efficiency.

- **Predictive Modeling:** AI algorithms can analyze historical data on fish growth, feeding patterns, and environmental conditions to predict optimal feeding times and quantities. This predictive capability ensures that fish receive the right amount of feed at the right time, enhancing growth and reducing waste.
- **Automated Decision-Making:** AI-driven systems can autonomously adjust feeding strategies based on real-time data. For example, if a drop in water temperature is detected, the system may reduce feed delivery, as fish may not be as active in cooler water.
- **Growth Monitoring:** By analyzing data from sensors and cameras, AI systems can track the growth of individual fish or groups, identify underperforming fish, and make targeted adjustments to feeding. This allows for more efficient use of feed and ensures that all fish receive adequate nutrition.

3.4. Internet of Things (IoT) and Cloud-Based Platforms

The Internet of Things (IoT) and cloud-based platforms enable the integration of various sensors, automated feeders, and data analytics tools into a cohesive precision feeding system. These platforms offer several key benefits:

- **Remote Monitoring:** IoT systems allow aquaculture operators to monitor feeding systems in real-time from remote locations. Data from sensors and feeders can be accessed via smartphones or computers, giving operators greater flexibility in managing feeding processes.
- **Data Storage and Analysis:** Cloud-based platforms provide centralized storage of historical data, allowing for long-term analysis of feeding practices, growth rates, and environmental conditions. This data can be used to identify trends, optimize feeding strategies, and improve farm management over time.
- **Remote Diagnostics:** IoT systems can detect malfunctions in feeding equipment, sensors, or other components and alert operators immediately. This ensures that issues are addressed promptly, minimizing disruptions to feeding and preventing feed loss.

4. Functional Benefits of Precision Feeding Systems

The adoption of precision feeding systems in aquaculture offers a multitude of functional benefits that significantly enhance operational efficiency, sustainability, and profitability. These benefits stem from the precise delivery of feed tailored to the specific needs of fish, leading to improved growth rates and healthier fish populations. Below, we delve into the key functional benefits of precision feeding systems:

4.1. Reduction of Feed Waste

One of the most significant advantages of precision feeding systems is their ability to minimize feed waste. Traditional feeding methods often result in excess feed being dispensed, leading to a substantial amount of uneaten feed sinking to the bottom of ponds or tanks. This waste not only represents a financial loss but also poses environmental risks.

- **Controlled Feed Delivery:** Precision feeding systems use real-time data to determine the optimal amount of feed required for each feeding event based on fish behavior and appetite. This allows for controlled and targeted feed delivery, significantly reducing the likelihood of uneaten feed.
- **Environmental Impact:** By minimizing feed waste, the risk of water contamination is lowered. Uneaten feed decomposes and

contributes to nutrient pollution, which can lead to harmful algal blooms and deteriorating water quality. Reducing feed waste enhances the sustainability of aquaculture operations and supports healthier ecosystems.

4.2. Improved Feed Conversion Ratios (FCR)

Feed conversion ratio (FCR) is a crucial metric in aquaculture that measures the efficiency with which fish convert feed into body mass. Precision feeding systems contribute to improved FCR through:

- **Optimal Feed Utilization:** By delivering the exact amount of feed tailored to the nutritional needs of fish at different growth stages, precision feeding systems enhance nutrient absorption and utilization. Fish receive the nutrients they require without the negative effects of overfeeding, leading to better growth performance.
- **Economic Benefits:** Improved FCR results in faster growth rates and lower feed costs. Farmers can achieve their production targets more efficiently, ultimately enhancing the profitability of aquaculture operations.

4.3. Enhanced Fish Growth and Health

Precision feeding systems significantly promote the growth and health of fish populations by providing tailored nutrition based on their specific needs:

- **Nutritional Targeting:** These systems allow for the precise delivery of essential nutrients, vitamins, and minerals, which are critical for optimal growth and immune function. Fish receive a balanced diet that supports their health and well-being, reducing the risk of nutritional deficiencies.
- **Stress Reduction:** By aligning feeding practices with fish behavior and environmental conditions, precision feeding systems help mitigate stress. Stress in fish can lead to increased susceptibility to diseases, lower growth rates, and higher mortality. By ensuring that fish are fed in a timely and appropriate manner, precision feeding helps maintain optimal health and resilience.

4.4. Environmental Sustainability

The environmental benefits of precision feeding systems extend beyond waste reduction. These systems play a vital role in promoting sustainable aquaculture practices:

- **Nutrient Management:** Precision feeding minimizes the introduction of excess nutrients into aquatic environments. In open-water aquaculture systems, this is particularly crucial, as nutrient overload can lead to harmful algal blooms and subsequent ecological imbalances. By controlling feed delivery and reducing waste, precision feeding supports better nutrient management.

- **Carbon Footprint Reduction:** With more efficient feed utilization and reduced waste, the overall carbon footprint of aquaculture operations can be minimized. This contributes to more sustainable food production practices that align with global goals for environmental conservation and sustainability.

4.5. Labor and Cost Efficiency

The automation and data-driven nature of precision feeding systems lead to significant labor and cost efficiencies in aquaculture operations:

- **Reduced Labor Requirements:** Automated feeding systems minimize the need for manual feeding, allowing operators to allocate human resources to other critical tasks. This not only increases operational efficiency but also reduces labor costs associated with routine feeding practices.
- **Cost-Effective Operations:** Data analytics and AI integration enable farmers to optimize feeding strategies based on real-time data. By ensuring that feed is utilized more effectively, aquaculture operations can lower overall feed expenditures. This leads to improved profitability and better resource allocation, further supporting sustainable practices.

5. Applications of Precision Feeding in Aquaculture Systems

Precision feeding systems have the potential to revolutionize various aquaculture practices by optimizing feeding strategies tailored to specific species, farming environments, and production goals. By leveraging advanced technologies, these systems can be effectively applied across different aquaculture systems, resulting in improved growth rates, reduced feed waste, and enhanced sustainability. The following sections detail the applications of precision feeding in key aquaculture settings:

5.1. Recirculating Aquaculture Systems (RAS)

Recirculating aquaculture systems (RAS) are designed to maintain optimal water quality by continuously filtering and reusing water. The integration of precision feeding systems in RAS offers numerous benefits:

- **Waste Reduction:** In RAS, the accumulation of uneaten feed and metabolic waste can quickly degrade water quality. Precision feeding systems minimize feed wastage by delivering precise amounts based on fish activity and appetite, which helps maintain cleaner water and reduces the frequency of water changes.

- **Optimized Feed Management:** By continuously monitoring fish behavior and environmental conditions, RAS can adapt feeding strategies in real-time. This ensures that fish receive the appropriate amount of nutrition while avoiding excess feed, contributing to healthier fish populations and improved growth rates.

5.2. Open-Water Aquaculture (Marine and Freshwater)

In open-water aquaculture systems, such as marine and freshwater cages, precision feeding systems provide significant advantages in managing large schools of fish:

- **Monitoring Tools:** Underwater cameras and acoustic sensors are utilized to observe fish feeding behavior and activity levels. This data is crucial for adjusting feed delivery in real-time, ensuring that all fish receive adequate nutrition without overfeeding.
- **Environmental Protection:** By optimizing feed delivery in open-water environments, precision feeding reduces the risk of excess feed sinking to the bottom and polluting the surrounding ecosystem. This is particularly important for minimizing nutrient loading in sensitive aquatic environments, which can lead to harmful algal blooms and other ecological disturbances.

5.3. Pond Aquaculture

Traditional pond aquaculture practices can also benefit from precision feeding systems:

- **Environmental Parameter Tracking:** Precision feeding technologies can monitor various environmental factors, such as water temperature, oxygen levels, and feed conversion rates. This information can be used to optimize feed delivery schedules and quantities based on real-time fish activity.
- **Improved Feed Efficiency:** By tailoring feed distribution to the specific needs of fish, precision feeding minimizes waste accumulation at the pond bottom. This leads to improved feed utilization and promotes a healthier pond ecosystem, supporting the growth of fish populations while reducing environmental impacts.

5.4. Intensive and Semi-Intensive Systems

In both intensive and semi-intensive aquaculture systems, where fish are stocked at high densities, precision feeding becomes essential:

- **Nutritional Delivery:** High stocking densities can complicate feeding practices and increase the risk of feed wastage. Precision feeding systems ensure that all fish receive the necessary nutrients without overloading the system with uneaten feed, which is critical for maintaining optimal growth and health.
- **Cost Management:** The efficient use of feed in these systems translates to significant cost savings for aquaculture operations. By improving feed conversion ratios and reducing waste, operators can achieve better economic returns and improve the overall sustainability of their practices.

6. Challenges and Future Directions

Despite the numerous benefits associated with precision feeding systems, several challenges must be addressed to facilitate their widespread adoption in aquaculture:

6.1. High Initial Costs

The initial investment required for precision feeding systems can be significant, including the costs of sensors, automated feeders, and data analytics platforms. While these costs can be justified by the long-term savings in feed expenses and improved productivity, the financial barrier can deter some operators from adopting these technologies.

6.2. Technological Integration

Integrating multiple sensors, feeders, and data platforms into a cohesive precision feeding system can be complex. Many aquaculture operators may lack the technical expertise required to effectively implement and manage these advanced systems. Developing user-friendly, robust solutions that simplify integration will be critical for encouraging broader adoption.

6.3. Data Management and Analysis

Precision feeding systems generate vast amounts of data, which can overwhelm operators without the appropriate tools and training. Effective data management and analysis are necessary to derive actionable insights from this information. Providing training programs for aquaculture operators to interpret data and make informed feeding decisions will be essential.

6.4. Species-Specific Adaptation

Different fish species exhibit unique feeding behaviors, nutritional requirements, and environmental tolerances. Precision feeding strategies must be tailored to the specific needs of various species to maximize their effectiveness. Further research and development are needed to optimize precision feeding systems for a wider range of fish species, including those that are less commonly farmed.

Table: Precision feeding systems in aquaculture, focusing on how they reduce waste and maximize growth

Aspect	**Description**	**Benefits**
Definition	Precision feeding systems deliver the right amount of feed at the right time, tailored to fish needs.	Optimizes feeding practices and minimizes waste.
Key Technologies	- Sensors (water quality, fish behavior) - Automated feeders - Data analytics - AI	Facilitates real-time monitoring and adjustments in feeding strategies.
Tailored Feeding	Adjusts feed based on species, growth stage, and environmental conditions.	Enhances nutrient utilization and promotes optimal growth.
Minimization of Feed Loss	Controlled feed delivery to prevent overfeeding and waste accumulation.	Reduces environmental pollution and improves water quality.
Nutrient Optimization	Nutritional formulations are designed to meet precise dietary needs.	Improves feed conversion ratios (FCR) and fish health.
Data-Driven Decisions	Continuous monitoring of fish behavior and environmental conditions guides feeding strategies.	Enables adaptive management and timely responses to changing conditions.
Environmental Impact	Reduces nutrient leaching and waste associated with overfeeding.	Supports sustainability and reduces ecological footprint.
Cost Efficiency	Reduces labor costs through automation and enhances profitability via optimized feed usage.	Maximizes economic returns on aquaculture investments.
Applications	- Recirculating Aquaculture Systems (RAS) - Open-water aquaculture - Pond aquaculture	Enhances performance in various farming environments.
Challenges	- High initial costs - Technological integration - Data management	Addressing these challenges can lead to widespread adoption and benefits.

7. Conclusion

Precision feeding systems present a transformative approach to aquaculture, enabling more efficient, sustainable, and data-driven management of feeding practices. By significantly reducing feed waste, improving feed conversion ratios, enhancing fish health, and minimizing environmental impact, these systems have the potential to revolutionize aquaculture operations. As technology continues to evolve and become more accessible, precision feeding is poised to play an increasingly important role in meeting the global demand for aquaculture products. The successful integration of precision feeding systems into diverse aquaculture settings will not only support the economic viability of the industry but also contribute to its long-term sustainability. By addressing the challenges and capitalizing on the opportunities presented by precision feeding, aquaculture can move towards a more resilient and sustainable future.

References

Barlow, J. W., & Leach, J. H. (2013). Use of technology to improve fish feeding practices: A review of current and emerging technologies. Aquaculture Research, 44(3), 1-10.

Barrows, F. T., & Hardy, R. W. (2002). Nutritional biotechnology in aquaculture. In Fish Nutrition (3rd ed., pp. 154-213). Academic Press.

Ewe, M. J., Hanel, R., & Wiegand, C. (2018). Precision feeding in aquaculture: The role of sensors. Aquaculture, 495, 636-645.

Figueiredo, H., & Azevedo, A. M. (2020). Precision aquaculture: Technologies, challenges, and opportunities. Aquaculture Reports, 18, 100482.

Gildberg, A., & Mørkøre, T. (2019). New developments in feeding technology for aquaculture. Fish Farming International, 8(1), 14-17.

He, J., Xu, X., & Li, Z. (2020). Application of precision feeding technology in aquaculture: A review. Journal of Aquaculture Research & Development, 11(6), 1-6.

Huang, J., Yang, J., & Wang, W. (2021). Intelligent feeding systems in aquaculture: Current applications and future perspectives. Aquacultural Engineering, 91, 102191.

Jørgensen, S. M., & Skov, P. V. (2018). Feed optimization in aquaculture: Challenges and solutions. Aquaculture International, 26(2), 285-300.

Kaushik, S. J., & Piatkowski, U. (2018). Precision feeding in aquaculture: A tool for sustainable production. Aquaculture Environment Interactions, 10, 251-261.

Kristensen, J. B., & Krogdahl, Å. (2018). Precision feeding: Optimizing feed delivery to farmed fish. Fish Physiology and Biochemistry, 44(1), 1-15.

Li, H., & Huang, W. (2020). Applications of precision aquaculture in fish farming: A review. Aquaculture, 520, 734947.

Martínez-Porchas, M., & Vargas-Albores, F. (2019). Innovations in aquaculture nutrition: Feeding strategies to reduce waste. Aquaculture Nutrition, 25(2), 235-246.

Nasir, U. R., & Awan, U. A. (2019). Potential of precision feeding technology in aquaculture: A review. Aquaculture Research, 50(6), 1581-1590.

Naylor, R. L., & Burke, M. (2021). Aquaculture: A global perspective on sustainability and waste. Global Environmental Change, 67, 102244.

Oguntola, J. A., & Olubunmi, M. (2020). The role of precision feeding in sustainable aquaculture. Sustainability, 12(14), 5628.

Pichavant, K., & Lahlou, A. (2020). Environmental impacts of fish farming and the role of feeding strategies: A review. Aquaculture Reports, 17, 100421.

Rahman, M. M., & Mazumder, N. H. (2021). Precision feeding: An innovative approach in aquaculture. International Journal of Fisheries and Aquatic Studies, 9(1), 20-26.

Rønnestad, I., & Krogdahl, Å. (2017). Precision feeding: A new paradigm for aquaculture nutrition. Aquaculture Research, 48(4), 1-11.

Sigernes, R., & Håstein, T. (2021). Smart feeding technology in aquaculture: Impacts on fish health and production efficiency. Aquaculture Reports, 19, 100552.

Van de Braak, K. M., & Van der Meer, J. (2019). Precision feeding in aquaculture: The future of feeding strategies. Aquaculture International, 27(3), 563-575.

4

Nutrient Recycling in Aquaculture from Waste to Feed

1. Introduction

Aquaculture has become one of the most significant contributors to global food production, responding to the escalating demand for fish and seafood. As a rapidly growing industry, aquaculture plays a crucial role in ensuring food security, nutrition, and livelihoods. According to the Food and Agriculture Organization (FAO), aquaculture production has expanded at an average annual rate of 5.3% over the past 30 years. This growth has positioned aquaculture as a primary source of protein, essential fatty acids, and micronutrients for billions of people worldwide. Despite its rapid expansion, the aquaculture industry faces significant environmental challenges, primarily due to the large amounts of organic waste generated by fish farming. This waste includes uneaten feed, fish excreta, and the remnants of dead organisms, all of which can contribute to nutrient pollution when not properly managed. Nutrient pollution is associated with serious ecological consequences such as eutrophication, where excess nutrients lead to algal blooms, oxygen depletion, and the degradation of aquatic habitats. These environmental issues threaten biodiversity and undermine the sustainability of aquaculture systems.

Nutrient recycling has emerged as an innovative and sustainable solution to address these challenges. This approach transforms waste into valuable resources, particularly in the form of nutrient-rich feed ingredients. By harnessing nutrients such as nitrogen, phosphorus, and organic matter from aquaculture waste, nutrient recycling not only reduces environmental impact but also improves the economic efficiency of aquaculture operations. Nutrient recycling promotes a circular economy model, where waste from fish farming is reintroduced into the system, closing nutrient loops and reducing the reliance on external inputs.

This chapter explores the principles, methods, benefits, and challenges of nutrient recycling in aquaculture. It emphasizes the transformative potential of nutrient recycling in creating a more sustainable and resilient aquaculture industry. The discussion will cover various nutrient recycling strategies, their implementation in aquaculture systems, and the technological innovations that enable waste management to be viewed as an opportunity for resource recovery.

2. Understanding Nutrient Recycling in Aquaculture

Nutrient recycling in aquaculture involves the recovery of essential nutrients from waste products generated during aquaculture operations. These recovered nutrients are reprocessed and reused in feed formulations, contributing to more sustainable and efficient production systems. This process plays a vital role in mitigating the environmental footprint of aquaculture, reducing nutrient pollution, and enhancing the economic sustainability of fish farming. The key components of nutrient recycling include the types of nutrients recovered, methods of recycling, and their integration into aquaculture systems.

2.1. Types of Nutrients Recycled

Aquaculture waste is rich in several nutrients essential for fish growth and health. The most significant nutrients that can be recycled from fish waste include nitrogen, phosphorus, and organic matter.

Nitrogen

Nitrogen is a fundamental nutrient in protein synthesis and growth for aquatic species. In aquaculture, nitrogen is derived from fish excreta, uneaten feed, and the breakdown of dead organisms. Unfortunately, only a fraction of nitrogen in fish feed is utilized for growth, with the remainder being excreted as ammonia or other nitrogenous compounds into the environment. Nitrogen recycling is crucial for reducing the environmental impact of nitrogen waste, which can contribute to harmful algal blooms and oxygen depletion in aquatic systems. One common method for nitrogen recycling is anaerobic digestion, a process in which organic waste is broken down by microorganisms in the absence of oxygen. This results in nitrogen-rich byproducts, such as ammonium, which can be used in the formulation of high-protein aquafeeds. Nitrogen recycling reduces the need for synthetic nitrogen fertilizers, lowering the cost of feed production and minimizing environmental harm.

Phosphorus

Phosphorus is another key nutrient necessary for cellular function, energy transfer, and skeletal development in fish. However, like nitrogen, a significant portion of the phosphorus in aquafeeds remains unused by the fish and is excreted in feces, contributing to nutrient pollution. Excess phosphorus can lead to eutrophication, which poses serious threats to the health of aquatic ecosystems. Recycling phosphorus from aquaculture waste can mitigate reliance on non-renewable phosphorus sources, such as phosphate rock, which are becoming increasingly scarce. Phosphorus recovery technologies can extract this valuable nutrient from fish waste, making it available for reuse in feed formulations or as a fertilizer for crops grown in integrated aquaculture systems. By recycling phosphorus, aquaculture can contribute to a more sustainable phosphorus cycle, reducing its environmental footprint.

Organic Matter

Organic matter from fish feces, uneaten feed, and dead organisms contributes significantly to aquaculture waste. Rich in carbon and nutrients, organic matter can be processed and recycled into nutrient-dense feed ingredients. Recycling organic matter enhances nutrient availability in feeds and supports the growth of beneficial microorganisms. Methods such as composting and fermentation can convert organic waste into a stable, nutrient-rich substrate. This substrate can then be used in the production of feed ingredients that are rich in proteins, carbohydrates, and essential nutrients. The recycling of organic matter also helps reduce the buildup of waste in aquaculture systems, improving water quality and the health of farmed species.

2.2. Mechanisms of Nutrient Recycling

The success of nutrient recycling in aquaculture depends on the efficient recovery and reprocessing of nutrients from waste. Several mechanisms facilitate nutrient recycling, ranging from biological processes to technological innovations.

Microbial Bioconversion

Microorganisms, including bacteria, fungi, and algae, play a crucial role in converting organic waste into bioavailable nutrients. These microbes decompose organic matter, breaking down complex compounds into simpler, nutrient-rich forms that can be reused in aquafeeds. Single-cell protein (SCP) derived from microbial conversion is an example of a protein-rich ingredient produced through this process. SCP is a sustainable alternative to traditional fishmeal, offering high protein content and essential amino acids for fish growth.

Integrated Multi-Trophic Aquaculture (IMTA)

IMTA is a system in which species from different trophic levels are cultivated together in the same environment. Waste from higher trophic species, such as fish, serves as nutrients for lower trophic species like shellfish and seaweed. These lower trophic organisms absorb dissolved nutrients, such as nitrogen and phosphorus, reducing nutrient pollution and converting waste into biomass that can be harvested and used as feed ingredients. IMTA exemplifies how nutrient recycling can be integrated into aquaculture systems to enhance sustainability and resource efficiency.

Insect-Based Recycling

Insects such as the black soldier fly (BSF) are increasingly used in aquaculture waste recycling. BSF larvae can consume large amounts of organic waste, converting it into a protein-rich biomass that can be processed into aquafeed. Insect-based recycling not only reduces waste but also provides a sustainable alternative to fishmeal, helping meet the growing demand for aquafeed ingredients while reducing pressure on wild fish stocks.

Algal Nutrient Recovery

Algae, particularly microalgae, can capture nutrients from aquaculture wastewater. These algae absorb excess nitrogen and phosphorus, reducing nutrient pollution while producing a biomass that is rich in protein, lipids, and essential fatty acids. The harvested algal biomass can then be processed into feed ingredients, contributing to both waste management and the production of high-quality feeds for farmed species.

3. Nutrient Recycling Methods

Nutrient recycling is a crucial component of sustainable aquaculture, allowing waste products from fish farming to be transformed into valuable resources. Various methods are employed to recycle nutrients, ranging from traditional approaches like composting to innovative techniques involving insects and algae. Each method offers distinct benefits and challenges, and their effectiveness depends on factors such as waste composition, system design, and the specific goals of the aquaculture operation.

3.1. Composting and Fermentation

Composting and fermentation are traditional methods of recycling organic waste into nutrient-rich substrates. These processes decompose organic matter from aquaculture systems, such as fish feces and uneaten feed, into humus-like materials that can be used in multiple ways.

- **Composting** involves aerobic decomposition, where microorganisms break down organic material in the presence of oxygen. This process converts waste into compost, which can be used as a soil amendment or as a fertilizer for plants in aquaponics systems. Composting also reduces the volume of waste, making it easier to manage.

- **Fermentation** is an anaerobic process that involves the breakdown of organic material by bacteria in the absence of oxygen. Fermented waste produces nutrient-dense substrates that can be added to fish feed as a supplement. Fermentation can also enhance the nutritional value of the waste by increasing its digestibility and bioavailability of certain nutrients.

Both composting and fermentation are relatively low-cost, making them accessible options for small-scale aquaculture operations looking to recycle waste into valuable feed ingredients or fertilizers. However, these methods require proper management to ensure that the process is efficient and that harmful pathogens or toxins are not present in the final product.

3.2. Anaerobic Digestion

Anaerobic digestion is a widely used method in nutrient recycling that breaks down organic matter in the absence of oxygen, producing biogas (methane) and a nutrient-rich byproduct known as digestate. This process is particularly effective for managing large volumes of organic waste generated by aquaculture systems, including fish excreta and uneaten feed.

- **Biogas production**: The methane generated through anaerobic digestion can be captured and used as a renewable energy source to power aquaculture operations, reducing reliance on external energy inputs.
- **Nutrient-rich digestate**: The solid and liquid fractions of the digestate contain high levels of nitrogen, phosphorus, and other micronutrients, which can be processed and used as fertilizers in aquaponics systems or as feed additives for fish. This approach helps reduce nutrient pollution and increases the overall efficiency of nutrient utilization in aquaculture.

Anaerobic digestion is particularly suitable for large-scale aquaculture systems due to its ability to handle significant waste volumes and its potential to generate both energy and valuable byproducts. However, the initial setup costs for anaerobic digestion systems can be high, and careful management is required to optimize methane production and ensure the safety of the resulting digestate for use in feed or agriculture.

3.3. Insect-Based Nutrient Recycling

Insect-based nutrient recycling is an innovative approach to waste management in aquaculture. One of the most commonly used insects for this purpose is the black soldier fly larvae (BSFL), which can consume large amounts of organic material, including fish waste, and convert it into protein-rich biomass.

- **Black Soldier Fly Larvae**: BSFL can be fed on a variety of organic wastes from aquaculture, such as fish feces, uneaten feed, and other organic detritus. The larvae rapidly break down the waste, converting it into nutrient-dense protein and fat, which can then be processed into high-quality feed for fish and other animals. BSFL meal is a sustainable alternative to traditional fishmeal, reducing the reliance on wild-caught fish stocks for aquafeed production.

Insect-based recycling offers numerous environmental and economic benefits. It not only reduces waste but also provides a renewable source of protein and fat for fish feed, thus reducing the need for costly and unsustainable fishmeal imports. This method is increasingly gaining attention as a solution for enhancing the sustainability of aquaculture systems while addressing global protein shortages.

3.4. Algae-Based Nutrient Recovery

Algal cultivation is another effective nutrient recycling method, particularly for capturing dissolved nitrogen and phosphorus from aquaculture effluents. Algae-based systems harness the natural ability of algae to absorb nutrients from water, promoting water quality improvement and biomass production.

- **Algal biomass**: The harvested algae contain high levels of proteins, lipids, and essential fatty acids, making them a valuable supplement for fish feed. Algal meal can enhance the nutritional quality of aquafeeds, providing a sustainable alternative to traditional ingredients such as fish oil and fishmeal.
- **Water quality improvement**: By absorbing excess nutrients from aquaculture systems, algae contribute to maintaining water quality, preventing issues like eutrophication. Algal systems can be integrated into recirculating aquaculture systems (RAS) to enhance nutrient recycling while supporting the growth of valuable microalgae species.

Algae-based nutrient recovery is a versatile and scalable method, suitable for both small- and large-scale aquaculture systems. However, the cultivation and harvesting of algae require careful management, and the processing of algal biomass into feed ingredients can involve additional costs.

3.5. Aquaponics Systems

Aquaponics is a sustainable, integrated farming system that combines aquaculture with hydroponics (growing plants in water without soil). In aquaponics, nutrient-rich wastewater from fish tanks is used to fertilize plants, which, in turn, absorb the nutrients and clean the water before it is recirculated back to the fish tanks.

- **Nutrient recycling**: In aquaponics, fish waste provides essential nutrients for plant growth, such as nitrogen and phosphorus. The plants remove these nutrients from the water, preventing their accumulation and potential harmful effects on the fish. This creates a closed-loop system where waste is continuously recycled, reducing the need for external inputs.
- **Simultaneous production of fish and crops**: Aquaponics allows for the production of both fish and vegetables or other crops in a single system, offering additional revenue streams and improving the overall economic sustainability of the farm.

Aquaponics systems are highly efficient in terms of resource use, as they recycle water and nutrients, reducing the environmental impact of both fish farming and agriculture. These systems are well-suited for areas with limited water resources or where sustainable farming practices are a priority. However, aquaponics systems can be complex to manage, requiring a balance between fish health, water quality, and plant nutrient requirements.

4. Benefits of Nutrient Recycling in Aquaculture

Nutrient recycling in aquaculture offers numerous environmental, economic, and operational benefits. By converting waste into valuable resources, aquaculture systems become more sustainable and economically viable, while reducing their impact on the surrounding environment.

4.1. Environmental Benefits

- **Reduced Eutrophication**: Nutrient recycling prevents excess nitrogen and phosphorus from being released into natural water bodies, reducing the risk of eutrophication. This is especially important in areas where nutrient pollution from fish farms has led to ecological imbalances, such as algal blooms and hypoxic "dead zones."
- **Waste Reduction**: By converting waste into useful products such as feed or fertilizer, nutrient recycling reduces the volume of waste that must be managed or treated. This minimizes the environmental burden of waste disposal and helps avoid the pollution of nearby ecosystems.
- **Improved Water Quality**: Systems that integrate algae or other biofilters improve water quality by removing excess nutrients from aquaculture effluents. This leads to a healthier environment for fish and other aquatic species, enhancing overall farm productivity.

4.2. Economic Benefits

- **Lower Feed Costs**: One of the most significant expenses in aquaculture is the cost of feed, especially fishmeal and other imported ingredients. By recycling nutrients into feed ingredients, aquaculture operations can reduce their reliance on costly external inputs, improving profitability.
- **Energy and Resource Efficiency**: Nutrient recycling systems, such as anaerobic digestion, produce renewable energy (biogas) and valuable byproducts (digestate) that can be used to further reduce operational costs. The optimized use of resources reduces overall farm inputs and improves the economic sustainability of the operation.
- **Product Diversification**: Integrated systems such as IMTA and aquaponics allow farmers to produce a variety of products, including fish, crops, and shellfish. This diversification offers additional revenue streams and reduces financial risk, making aquaculture more resilient to market fluctuations.

4.3. Circular Economy

Nutrient recycling supports the concept of a circular economy in aquaculture, where waste is not discarded but repurposed into new resources. By closing the loop between waste generation and resource production, nutrient recycling enhances both environmental and economic sustainability.

5. Challenges in Nutrient Recycling

While nutrient recycling in aquaculture offers significant advantages in terms of sustainability and economic efficiency, there are still several challenges that must be addressed to fully realize its potential. These challenges encompass technical, regulatory, financial, and biological factors that affect the successful implementation of nutrient recycling systems in aquaculture operations.

5.1. Technical Barriers

One of the primary challenges in nutrient recycling is the technical complexity involved in integrating these systems into existing aquaculture operations, especially for small-scale farmers. Advanced nutrient recycling technologies, such as anaerobic digesters, algal bioreactors, or insect-based systems, require specific infrastructure and expertise that many farmers may not have access to.

- **System integration**: Many aquaculture systems are not designed with nutrient recycling in mind. Retrofitting these systems to accommodate nutrient recycling technologies can be both costly and technically challenging. For example, the incorporation of sensors, control systems, and specialized equipment to monitor nutrient levels, water quality, and waste management requires significant modifications.

- **Operational challenges**: Even after installation, maintaining optimal conditions for recycling, such as ensuring correct temperature, pH, and oxygen levels in composting or anaerobic digestion systems, can be difficult. Without proper management, nutrient recycling processes can fail, resulting in poor feed quality or environmental hazards.
- **Data and monitoring**: Advanced nutrient recycling systems require continuous monitoring and optimization to function efficiently. For instance, monitoring nutrient levels in aquaponics or algae-based recovery systems is essential to prevent nutrient imbalances. The lack of data management tools and automation in many aquaculture systems presents another barrier to the widespread adoption of nutrient recycling.

Overcoming these technical barriers will require a concerted effort to develop affordable, user-friendly, and scalable recycling technologies that can be implemented in diverse aquaculture systems, including small-scale operations.

5.2. Regulatory and Health Concerns

Another major challenge in nutrient recycling relates to the safety and regulatory frameworks governing the use of recycled nutrients in aquafeeds. The use of recycled waste products as feed ingredients raises concerns about potential contamination, bioaccumulation of harmful substances, and risks to both fish health and human consumers.

- **Feed safety**: When using recycled nutrients, such as composted waste or insect meal derived from aquaculture waste, there is a risk of contamination from pathogens, toxins, or heavy metals. Ensuring the safety of recycled feed ingredients requires stringent processing protocols, risk assessments, and quality control measures to prevent any harmful substances from entering the food chain.
- **Regulatory hurdles**: In many regions, regulatory frameworks governing the use of recycled feed ingredients in aquaculture are either underdeveloped or highly restrictive. For example, using insect-based proteins or algae-derived nutrients in feed may require approval from food safety authorities, which can involve lengthy and costly compliance processes.
- **Public perception**: There are also concerns regarding consumer acceptance of products derived from recycled waste, especially when it comes to food safety and environmental impacts. Transparency, public education, and clear labeling can help build trust in nutrient recycling

practices and products, but these measures will require time and coordinated efforts.

To address these challenges, clear and scientifically backed regulatory guidelines are needed. Collaboration between governments, research institutions, and the aquaculture industry can help establish these frameworks while ensuring the safety and sustainability of nutrient recycling.

5.3. Species-Specific Requirements

Aquaculture involves a diverse range of species, each with its own specific nutritional needs. One of the difficulties in using recycled feed ingredients is ensuring that these recycled products meet the dietary requirements of different fish species, which vary in terms of protein content, amino acid profile, lipid requirements, and micronutrient needs.

- **Nutrient variability**: Recycled feed ingredients, such as insect meal or algae, may not always provide consistent or sufficient levels of essential nutrients. For example, certain species may require high levels of omega-3 fatty acids or specific amino acids that may not be adequately supplied by recycled feeds. Inadequate nutrition can result in poor growth, compromised immune function, or other health issues in fish.
- **Tailored formulations**: More research is needed to develop species-specific formulations for recycled feeds. This involves determining the ideal balance of nutrients in recycled ingredients, understanding how different species metabolize these nutrients, and ensuring that the recycled feed does not negatively affect fish health or product quality.

Achieving species-specific nutrient formulations requires collaboration between nutritionists, feed manufacturers, and aquaculture producers. Continued research and innovation in feed science are essential for developing tailored, nutritionally balanced feeds from recycled nutrients.

5.4. Cost of Implementation

While nutrient recycling offers long-term benefits, the initial capital investment required to implement these systems can be prohibitively high, particularly for small-scale farmers. For instance, installing anaerobic digesters, algae cultivation systems, or insect rearing facilities involves significant upfront costs, and the return on investment may take several years to materialize.

- **High setup costs**: Advanced nutrient recycling technologies often require specialized equipment, infrastructure, and expertise, all of which contribute to high installation costs. Small and medium-scale farmers may lack the financial resources to make these investments without external support.

- **Operational costs**: In addition to setup costs, nutrient recycling systems require ongoing maintenance and management, which can add to the operational costs of aquaculture operations. Without government subsidies, financial incentives, or cost-sharing mechanisms, the adoption of nutrient recycling systems may remain limited.

Governments, international organizations, and the private sector can play a crucial role in overcoming these financial barriers by providing grants, subsidies, and financial incentives for adopting nutrient recycling technologies. Encouraging public-private partnerships and fostering innovation can also help lower the costs associated with these systems.

6. Future Directions and Innovations

The future of nutrient recycling in aquaculture is highly promising, with several cutting-edge technologies and innovations that could enhance its feasibility and efficiency. Advances in bioprocessing, genetic engineering, and precision farming are expected to play a pivotal role in overcoming the current challenges and scaling up nutrient recycling.

- **Microbial bioprocessing**: Advances in microbial technology, such as genetically engineered bacteria and fungi, could allow for more efficient conversion of organic waste into high-quality feed ingredients. These microbes could be tailored to break down specific waste compounds, enhance nutrient availability, or even produce bioactive compounds that boost fish growth and immunity.
- **Genetic engineering of algae and insects**: The genetic modification of algae and insects could improve their nutritional profiles, making them more suitable for use as feed ingredients. For example, algae strains could be engineered to produce higher levels of omega-3 fatty acids or essential amino acids, while insects could be bred for faster growth and higher protein yields.
- **Precision aquaculture**: The integration of precision farming technologies, such as sensors, automation, and data analytics, can optimize nutrient recycling processes by providing real-time information on water quality, nutrient levels, and fish health. These technologies can help farmers monitor and manage their recycling systems more effectively, reducing waste and improving feed efficiency.
- **Circular economy models**: Future innovations in aquaculture are expected to embrace circular economy principles, where waste is continually recycled and repurposed. This could include integrating aquaculture with other food production systems, such as agriculture or livestock farming, to create fully closed-loop systems.

In addition to technological advancements, collaborative efforts between researchers, industry stakeholders, and policymakers will be essential in developing regulatory frameworks that support nutrient recycling while ensuring food safety and environmental sustainability.

7. Conclusion

Nutrient recycling in aquaculture offers a transformative approach to addressing the industry's environmental challenges while improving economic sustainability. By converting waste into feed, nutrient recycling reduces pollution, optimizes resource use, and lowers feed costs. However, for nutrient recycling to reach its full potential, key challenges such as technical integration, regulatory compliance, species-specific nutrition, and cost barriers must be addressed. As innovations in microbial bioprocessing, genetic engineering, and precision farming continue to evolve, nutrient recycling is expected to become a cornerstone of sustainable aquaculture. With appropriate regulatory frameworks and industry collaboration, the aquaculture sector can transition toward a circular economy model, ensuring that waste is no longer a burden but a valuable resource for future food production.

References

Barrows, F. T., Gaylord, T. G., Sealey, W. M., Smith, C. E., & Porter, L. (2010). The effect of different dietary levels of protein, lipid, and carbohydrate on growth and feed efficiency of rainbow trout (Oncorhynchus mykiss). Aquaculture, 305(1-4), 182-188.

Bossier, P., & Ekasari, J. (2017). Biofloc technology application in aquaculture to support sustainable development goals. Microbial biotechnology, 10(5), 1012-1016.

Cao, L., Diana, J. S., & Keoleian, G. A. (2013). Aquaculture systems: environmental impact and resource use. Annual Review of Environment and Resources, 38, 349-372.

Dhont, J., Dierckens, K., Stappen, G.V., & Sorgeloos, P. (2016). Aquatic feed resources: Feed production and utilization. World Aquaculture Society.

Ghaly, A. E., & MacDonald, K. N. (2010). Biological wastewater treatment in recirculating aquaculture systems: A review. Aquaculture and Fisheries Management, 10, 126-135.

Goddek, S., Delaide, B., Mankasingh, U., Ragnarsdóttir, K. V., Jijakli, M. H., & Thorarinsdottir, R. (2015). Challenges of sustainable and commercial aquaponics. Sustainability, 7(4), 4199-4224.

Karak, T., Bhagat, R. M., & Bhattacharyya, P. (2013). Municipal solid waste generation, composition, and management: The world scenario. Critical Reviews in Environmental Science and Technology, 42(15), 1509-1630.

Naylor, R. L., Hardy, R. W., Bureau, D. P., Chiu, A., Elliott, M., Farrell, A. P., & Nichols, P. D. (2009). Feeding aquaculture in an era of finite resources. Proceedings of the National Academy of Sciences, 106(36), 15103-15110.

Ricker, R. W. (2014). Microbial processing of aquaculture effluent for nutrient recovery and resource conservation. Journal of Aquaculture Research and Development, 5(6), 321-330.

Sarker, P. K., Kapuscinski, A. R., Lanois, A., Livesey, E. D., Bernhard, K. P., & Coley, M. L. (2016). Towards sustainable aquafeeds: Complete substitution of fish oil with marine microalga Schizochytrium sp. improves growth and fatty acid deposition in juvenile Nile tilapia (Oreochromis niloticus). PLoS one, 11(6), e0156684.

Schmitt, V., & Lopez, C. A. (2017). Integrated multitrophic aquaculture: a tool to mitigate environmental impacts. Aquaculture Environment Interactions, 9, 235-246.

Stankus, A. (2013). Commercialization of black soldier fly larvae meal: an alternative protein source in aquaculture feed. Journal of Applied Aquaculture, 25(4), 321-340.

Tacon, A. G., Hasan, M. R., & Metian, M. (2011). Demand and supply of feed ingredients for farmed fish and crustaceans: Trends and prospects. FAO Fisheries and Aquaculture Technical Paper No. 564.

Timmons, M. B., & Ebeling, J. M. (2013). Recirculating aquaculture. Cayuga Aqua Ventures, 2nd edition.

Vallejo, A., Álvarez, J. A., & Barrena, R. (2017). Nutrient recovery from aquaculture wastewater using constructed wetlands. Environmental Technology, 38(20), 2710-2721.

Van Rijn, J., Tal, Y., & Schreier, H. J. (2006). Denitrification in recirculating aquaculture systems: From biochemistry to biofilters. Aquaculture, 278(1-4), 1-14.

Verdegem, M. C., Bosma, R. H., & Verreth, J. A. (2006). Reducing water use for animal production through aquaculture. International Journal of Water Resources Development, 22(1), 101-113.

Wang, J., Yang, C., & Zhang, X. (2017). Nutrient removal from aquaculture wastewater using a microalgal biofilm system. Bioresource Technology, 241, 404-411.

Ytrestøyl, T., Aas, T. S., & Åsgård, T. (2015). Utilisation of feed resources in production of Atlantic salmon (Salmo salar) in Norway. Aquaculture, 448, 365-374.

Zhang, D., Luo, G., Tan, H., & Chen, L. (2020). Nutrient recovery from aquaculture waste: Applications of algae-based systems. Bioresource Technology Reports, 12, 100582.

5

Phytogenics Plant-Based Additives for Healthier Fish

1. Introduction

Aquaculture, the farming of aquatic organisms such as fish, shellfish, and aquatic plants, has become one of the fastest-growing food production sectors globally. This growth has been driven by the increasing demand for high-quality protein to feed a growing global population, coupled with the depletion of wild fish stocks due to overfishing. As the aquaculture industry expands, so too does the need for sustainable practices that balance productivity with environmental and animal welfare considerations. One of the critical challenges faced by modern aquaculture is the need to maintain fish health and growth performance while minimizing the industry's ecological footprint. Traditionally, synthetic additives such as antibiotics, chemical growth promoters, and other artificial substances have been used to improve fish health, enhance growth rates, and prevent disease outbreaks. However, the excessive use of these chemicals has led to several issues, including antibiotic resistance, chemical residues in fish products, and environmental pollution. These concerns have spurred a shift toward more sustainable, eco-friendly alternatives in aquaculture production systems.

In this context, phytogenics—natural, plant-based additives—have garnered increasing interest as a promising solution. Phytogenics are bioactive compounds extracted from various plant sources, including herbs, spices, and essential oils, that possess multiple beneficial properties. These compounds have been used in traditional medicine and animal husbandry for centuries due to their antimicrobial, antioxidant, anti-inflammatory, and immunomodulatory effects. By leveraging the natural bioactive compounds in plants, phytogenics offer a range of advantages for aquaculture, including enhanced growth performance, improved feed efficiency, strengthened immune responses, and reduced disease susceptibility. The need for sustainable and natural alternatives to synthetic additives is becoming more urgent as the global consumer base becomes more health-conscious and environmentally aware. Consumers today increasingly demand seafood products that are free from chemical additives and produced through environmentally responsible methods. Consequently, the aquaculture industry must adapt by incorporating more natural and sustainable approaches to meet these expectations. This shift is also supported by stringent regulations in many countries that are restricting the use of antibiotics and synthetic chemicals in aquaculture production.

Growing Need for Sustainable Solutions in Aquaculture

As the global population continues to rise, the demand for fish as a primary protein source is projected to grow exponentially in the coming decades. By 2050, it is estimated that the world will need to produce nearly 60% more seafood than current levels to meet this demand. To achieve this, aquaculture must adopt innovative and sustainable practices to maintain fish production without compromising environmental integrity or animal health. Sustainability in aquaculture is not just about reducing the industry's environmental impact; it also involves enhancing the nutritional value of the feed, improving the welfare of farmed fish, and ensuring the long-term viability of production systems. Phytogenics offer a solution that aligns with these goals by promoting healthier fish, enhancing feed conversion efficiency, and reducing reliance on artificial additives.

Concerns over Antibiotic Use in Aquaculture

One of the primary drivers behind the search for alternatives like phytogenics is the growing concern over the use of antibiotics in aquaculture. Antibiotics have been widely used to prevent and treat bacterial infections in fish, as well as to promote growth. However, the misuse and overuse of antibiotics have led to the development of antibiotic-resistant bacteria, posing significant risks to both animal and human health. Resistant strains of bacteria can transfer from fish to humans through the food chain, contributing to the global public health crisis of antimicrobial resistance. Moreover, the residual presence of antibiotics in water bodies and fish products raises serious environmental and consumer health concerns. Many countries have implemented stricter regulations and guidelines to limit the use of antibiotics in aquaculture. In response, the industry is exploring natural alternatives like phytogenics, which can enhance fish health and immunity without the associated risks of antibiotics.

Phytogenics: A Natural Alternative for Sustainable Aquaculture

Phytogenics have emerged as a potent, natural alternative to synthetic additives and antibiotics in aquaculture systems. These plant-derived compounds offer multiple benefits without the adverse environmental or health impacts associated with conventional chemical inputs. By harnessing the natural properties of herbs, spices, and essential oils, phytogenics improve growth performance, immune function, feed efficiency, and overall fish welfare. Plant-based additives such as garlic (Allium sativum), ginger (Zingiber officinale), oregano (Origanum vulgare), and turmeric (Curcuma longa) contain bioactive compounds like alkaloids, flavonoids, and terpenoids that have been shown to possess antimicrobial, antioxidant, and immunomodulatory effects. These compounds help prevent infections, reduce oxidative stress, and enhance the resilience of fish to various stressors. The shift towards phytogenics aligns with the broader global movement toward sustainable and organic food production. By integrating phytogenic additives into aquaculture feeds, the industry can reduce its reliance on synthetic chemicals, improve the health and welfare of farmed fish, and minimize environmental pollution. Moreover, phytogenics can help meet consumer demand for healthier, chemical-free seafood, while contributing to more sustainable aquaculture systems.

The Scope of Phytogenics in Aquaculture

The use of phytogenics in aquaculture is not limited to one specific benefit; rather, it offers a holistic approach to improving fish health and production efficiency. These plant-based additives can serve as growth promoters, immune enhancers, antimicrobial agents, and stress mitigators, making them versatile tools for aquaculture producers. Moreover, the wide variety of plant compounds available allows for the formulation of specific additives tailored to the needs of different fish species and production systems. This chapter aims to explore the role of phytogenics in aquaculture, examining their modes of action, specific plant-based additives used in fish farming, and the evidence supporting their efficacy. The chapter will also discuss the challenges and future potential of phytogenics, highlighting how they can contribute to the development of healthier, more sustainable aquaculture practices. Through this exploration, we will see how phytogenics not only enhance the productivity of aquaculture operations but also contribute to the sustainability and ethical responsibility of the industry. As the demand for environmentally friendly and natural production methods grows, phytogenics are poised to become a cornerstone of modern, sustainable aquaculture.

2. What are Phytogenics?

Phytogenics, also referred to as phytobiotics or plant-based additives, encompass a broad category of natural bioactive compounds derived from plants, including herbs, spices, and essential oils. These compounds are increasingly being integrated into aquafeeds to improve the growth, health, and overall performance of aquatic species. Derived from nature's diverse plant kingdom, phytogenics include a wide array of bioactive substances such as flavonoids, alkaloids, terpenoids, phenolics, tannins, and saponins, each offering distinct biological benefits.

Phytogenic additives are celebrated for their multifaceted properties, including antimicrobial, antioxidant, anti-inflammatory, and immunomodulatory effects. These attributes make them an attractive alternative to conventional synthetic additives, such as antibiotics and chemical growth promoters, which have been associated with negative environmental and health impacts. The use of phytogenics in aquaculture is particularly appealing due to the natural, sustainable, and non-toxic nature of these compounds.

Phytogenics can be classified into several categories based on their origin and method of extraction:

- **Herbs**: Whole or ground parts of plants like leaves, stems, or roots. Common herbs used in aquaculture include oregano (*Origanum vulgare*), thyme (*Thymus vulgaris*), and rosemary (*Rosmarinus officinalis*).
- **Spices**: Parts of plants rich in bioactive compounds, such as turmeric (*Curcuma longa*), garlic (*Allium sativum*), and black pepper (*Piper nigrum*), are widely used in aquafeeds.
- **Essential oils**: Concentrated volatile oils extracted from various plants. These oils, often rich in terpenoids and phenolic compounds, exhibit strong antimicrobial, antioxidant, and digestive-stimulant properties. Examples include oils from oregano, thyme, and clove (*Syzygium aromaticum*).

The inclusion of phytogenics in aquaculture feeds provides an array of benefits that align with the industry's growing focus on sustainability. These plant-based additives are recognized for their ability to enhance growth performance, improve feed conversion, boost immune responses, and protect fish from diseases. By offering natural, plant-derived solutions, phytogenics hold significant potential for advancing healthier and more sustainable aquaculture practices.

3. Benefits of Phytogenics in Aquaculture

The introduction of phytogenic additives into aquafeeds has garnered significant attention due to the multitude of benefits they offer for fish health, growth, and sustainability. Below, we delve into some of the key benefits of phytogenics in aquaculture:

3.1. Growth Promotion and Feed Efficiency

One of the primary benefits of phytogenics is their ability to promote growth and enhance feed conversion efficiency (FCR). Phytogenics often improve the palatability of feed, stimulating appetite and encouraging fish to consume more feed. Certain compounds, such as alkaloids and terpenoids, also enhance digestive enzyme activity, resulting in better nutrient absorption and utilization. For instance, research has demonstrated that the inclusion of garlic (*Allium sativum*) and ginger (*Zingiber officinale*) extracts in the diet of Nile tilapia (*Oreochromis niloticus*) significantly improved weight gain and feed conversion ratios (FCR). These plants contain bioactive compounds that stimulate digestive enzyme production and improve gut health, thereby optimizing nutrient utilization. In addition to enhancing nutrient absorption, phytogenics can reduce the energy expended by fish during digestion, allowing more energy to be allocated towards growth. This improvement in feed efficiency directly translates to economic savings for aquaculture producers, as less feed is required to achieve desired growth rates.

3.2. Immune System Enhancement

Fish in aquaculture systems are often exposed to stressors such as overcrowding, poor water quality, and handling, all of which can compromise their immune systems. Phytogenics play a vital role in supporting the immune system, helping fish combat disease and maintain optimal health. Many plant-based compounds possess immunomodulatory properties, which can enhance the fish's innate and adaptive immune responses. For example, extracts from *Echinacea purpurea* and *Aloe vera* have been shown to boost the immune systems of carp (*Cyprinus carpio*) and catfish (*Clarias gariepinus*), respectively. These plants enhance the production of immune cells, such as white blood cells and macrophages, that are critical for fighting off pathogens. Phytogenics also stimulate the production of antibodies and other immune factors, providing fish with enhanced protection against bacterial, viral, and parasitic infections. By strengthening the fish's immune defenses, phytogenics reduce the need for antibiotics and other synthetic disease control measures, contributing to more sustainable and environmentally friendly aquaculture practices.

3.3. Antimicrobial Properties

Disease outbreaks caused by bacteria, fungi, and parasites are a major threat to aquaculture productivity. One of the most valuable attributes of phytogenics is their antimicrobial activity, which can help protect fish from harmful pathogens. Many phytogenic compounds, particularly those found in essential oils, possess potent antimicrobial properties that can inhibit the growth of bacteria, fungi, and even parasites. For instance, essential oils from oregano (*Origanum vulgare*), thyme (*Thymus vulgaris*), and clove (*Syzygium aromaticum*) contain active compounds like carvacrol, thymol, and eugenol, which have been proven to exhibit strong antimicrobial effects. These compounds disrupt the cell membranes of pathogens, preventing them from multiplying and causing disease in fish. By using phytogenics with antimicrobial properties, aquaculture operators can reduce their reliance on synthetic antibiotics, thereby minimizing the risk of antibiotic resistance and improving the sustainability of the industry. The use of phytogenics also contributes to safer seafood products, as there is less risk of antibiotic residues in the fish.

3.4. Antioxidant and Anti-Inflammatory Effects

Fish are regularly exposed to environmental stressors, such as temperature fluctuations, suboptimal water quality, and stocking density, which can lead to oxidative stress and inflammation. Phytogenics contain powerful antioxidant compounds that help fish combat oxidative stress by neutralizing free radicals. These antioxidants, such as polyphenols, flavonoids, and tannins, protect fish cells from damage and reduce the risk of inflammation and tissue degradation. For example, extracts from green tea (*Camellia sinensis*) and turmeric (*Curcuma longa*) are rich in polyphenols, which have been shown to improve antioxidant capacity in fish. These compounds not only reduce oxidative damage but also exhibit anti-inflammatory effects by inhibiting pro-inflammatory signaling pathways. This dual action helps fish maintain their health, especially during stressful conditions. The anti-inflammatory properties of phytogenics are particularly valuable in aquaculture, as chronic inflammation can negatively impact growth, immune function, and overall performance. By reducing inflammation, phytogenics help fish recover from environmental stressors more quickly and maintain better health and growth rates.

3.5. Gut Health Improvement

The digestive system is integral to fish health and growth, and the use of phytogenics can significantly enhance gut health. Many phytogenic compounds, such as tannins and saponins, act as prebiotics, promoting the growth of beneficial gut bacteria. These beneficial microbes, in turn, help to outcompete harmful pathogens and support the overall health of the gut microbiome. For example, garlic and ginger extracts have been shown to promote the development of a healthy gut environment by enhancing the production of digestive enzymes and improving the integrity of the intestinal mucosa. A healthy gut is essential for efficient nutrient absorption, as well as for protecting fish from enteric diseases. Phytogenics can also improve gut barrier function, reducing the likelihood of harmful pathogens penetrating the gut lining and causing systemic infections. As a result, fish that are fed phytogenic-enhanced diets are more likely to maintain optimal health and exhibit better growth performance.

Phytogenics represent a promising, natural solution for improving the health, growth, and overall performance of fish in aquaculture systems. These plant-based additives offer a range of benefits, from enhancing immune responses and promoting gut health to providing antimicrobial, antioxidant, and anti-inflammatory effects. By integrating phytogenics into aquafeeds, the aquaculture industry can move towards more sustainable, efficient, and environmentally friendly practices. As research on phytogenics continues to grow, their potential to transform aquaculture is becoming increasingly clear, offering a valuable tool for ensuring the long-term health and productivity of fish farms.

4. Common Phytogenics Used in Aquaculture

Phytogenics, derived from various plants, have gained considerable attention in aquaculture for their health-promoting, antimicrobial, and growth-enhancing properties. Each phytogenic source has unique bioactive compounds that contribute to specific benefits for fish health and performance. Here are some of the most commonly used phytogenics in aquaculture:

4.1. Garlic (*Allium sativum*)

Garlic is a powerful phytogenic widely used in aquaculture for its diverse health benefits. Its primary bioactive compound, allicin, exhibits potent antimicrobial properties that combat bacterial, viral, and parasitic infections in fish. Besides, sulfur-containing compounds in garlic stimulate the immune system, improving fish resistance to diseases.

- **Antimicrobial properties**: Allicin and other sulfur-containing compounds in garlic can suppress the growth of harmful bacteria like *Aeromonas* and *Vibrio* species, commonly responsible for infections in aquaculture.
- **Growth promotion**: Garlic is known to improve feed intake and digestive enzyme activity, resulting in enhanced feed efficiency and growth performance. In species such as tilapia (*Oreochromis niloticus*), garlic supplementation has led to significant improvements in weight gain and feed conversion ratios.
- **Immunostimulatory effects**: Garlic strengthens fish immunity by increasing the production of immune cells like macrophages and lymphocytes, enabling fish to better resist pathogens. This immune enhancement reduces mortality rates and improves overall health.

4.2. Ginger (*Zingiber officinale*)

Ginger is another popular phytogenic used in aquaculture, known for its strong anti-inflammatory, antioxidant, and immune-boosting properties. Rich in bioactive compounds like gingerol and shogaol, ginger contributes to multiple health benefits for fish.

- **Anti-inflammatory and antioxidant effects**: Ginger can reduce inflammation and oxidative stress in fish by neutralizing harmful free radicals. This helps in maintaining fish health under stressful conditions, such as poor water quality or high stocking densities.
- **Digestive health**: Ginger promotes digestive enzyme activity, which enhances nutrient absorption and improves feed conversion. Studies have shown that ginger-supplemented diets lead to better growth performance in fish species like catfish (*Clarias gariepinus*) and carp (*Cyprinus carpio*).
- **Immune enhancement**: Ginger strengthens the fish immune system by boosting the production of white blood cells, improving resistance to infections from pathogens like *Aeromonas* and *Pseudomonas*.

4.3. Oregano (*Origanum vulgare*)

Oregano is well-regarded for its high concentrations of carvacrol and thymol, two compounds with potent antimicrobial effects. These compounds make oregano a valuable phytogenic for preventing bacterial and parasitic infections in aquaculture systems.

- **Antimicrobial activity**: Oregano effectively inhibits the growth of harmful pathogens like *Vibrio*, *Aeromonas*, and *Streptococcus* species. Its antimicrobial properties are attributed to the disruption of bacterial cell membranes, leading to cell death.
- **Growth performance**: Oregano has been shown to improve growth performance and feed efficiency in several fish species by enhancing nutrient absorption and gut health. The use of oregano essential oil in aquafeeds for European seabass (*Dicentrarchus labrax*) has resulted in improved weight gain and better overall fish performance.

4.4. Turmeric (*Curcuma longa*)

Turmeric contains curcumin, a well-known compound with powerful antioxidant and anti-inflammatory properties. Its ability to reduce oxidative stress and inflammation makes it a valuable phytogenic for maintaining fish health.

- **Antioxidant effects**: Curcumin neutralizes free radicals and prevents oxidative damage to fish tissues, which is particularly beneficial during stressful conditions like heat stress or disease outbreaks.
- **Immune enhancement**: Turmeric enhances immune responses by stimulating the production of immune cells and increasing the activity of natural killer cells, helping fish fight off infections.
- **Liver health**: Curcumin has hepatoprotective properties, improving liver function and promoting detoxification in fish. This is especially important in aquaculture systems where fish are often exposed to various environmental pollutants and toxins.

4.5. Aloe Vera (*Aloe barbadensis*)

Aloe vera is known for its soothing, healing, and immune-boosting effects, making it a useful phytogenic for aquaculture applications. It contains bioactive compounds such as polysaccharides and glycoproteins that support immune health and wound healing in fish.

- **Immune stimulation**: Aloe vera extracts stimulate immune responses, enhancing the production of antibodies and immune cells. This helps fish resist infections from pathogens such as *Aeromonas hydrophila* and *Vibrio parahaemolyticus*.
- **Wound healing**: Aloe vera has been used to promote wound healing in fish, especially in species prone to external injuries or skin damage. Its bioactive compounds accelerate the repair of damaged tissues and improve overall skin health.
- **Growth enhancement**: Aloe vera supplementation in aquafeeds has been linked to improved growth rates and feed efficiency in fish species like catfish and tilapia.

5. Challenges and Considerations

While phytogenics offer numerous benefits, there are several challenges that must be addressed to maximize their efficacy in aquaculture:

- **Variability in quality**: The quality and efficacy of phytogenics can vary depending on factors such as the source of the plant, extraction methods, and storage conditions. Inconsistent quality can affect the reliability of their effects on fish health and performance.
- **Bioavailability**: The bioavailability of active compounds in phytogenics can be limited by factors such as digestion and metabolism in fish. Finding ways to enhance the absorption of these compounds is crucial for ensuring their effectiveness.

- **Interactions with feed ingredients**: Phytogenics may interact with other feed components, potentially influencing their effects on fish. More research is needed to understand these interactions and how they impact the overall formulation of aquafeeds.
- **Cost**: The cost of high-quality phytogenics can be a limiting factor, especially in comparison to cheaper synthetic additives. However, the long-term benefits of improved fish health and reduced antibiotic use may offset the initial investment.

6. Future Directions and Innovations

The use of phytogenics in aquaculture is expected to grow as the industry focuses on sustainable and natural solutions for enhancing fish health and productivity. Several innovations are being explored to improve the efficacy and cost-effectiveness of phytogenics:

- **Advanced extraction technologies**: New methods for extracting and preserving bioactive compounds from plants are being developed, which can increase the potency and consistency of phytogenics.
- **Synergistic effects**: Research into the synergistic interactions between different phytogenic compounds could lead to the development of more effective formulations that provide enhanced health benefits.
- **Novel delivery systems**: Innovations in encapsulation and other delivery technologies may improve the bioavailability of phytogenics, ensuring that active compounds are efficiently absorbed by fish.

Table: Common Phytogenics and Their Benefits in Aquaculture

Phytogenic Additive	Source Plant	Active Compounds	Primary Benefits	Fish Species
Garlic (Allium sativum)	Garlic	Allicin, sulfur compounds	Antimicrobial, immune boost, growth promotion	Tilapia, Catfish, Salmonids
Ginger (Zingiber officinale)	Ginger	Gingerol, shogaol	Anti-inflammatory, antioxidant, improves feed efficiency	Nile tilapia, Rainbow trout, Carp
Oregano (Origanum vulgare)	Oregano	Carvacrol, thymol	Antimicrobial, antiparasitic, enhances feed efficiency	Salmon, Carp, Tilapia
Turmeric (Curcuma longa)	Turmeric	Curcumin	Antioxidant, anti-inflammatory, improves liver function	Tilapia, Carp, Catfish

Aloe Vera (Aloe barbadensis)	Aloe Vera	Polysaccharides, vitamins	Immune health, wound healing, antimicrobial properties	Carp, Catfish, Shrimp
Thyme (Thymus vulgaris)	Thyme	Thymol, carvacrol	Antimicrobial, improves digestion, antioxidant	Tilapia, Salmon, Trout
Echinacea (Echinacea purpurea)	Echinacea	Polysaccharides, phenolics	Immunostimulant, antibacterial, stress reduction	Carp, Catfish, Shrimp
Green Tea (Camellia sinensis)	Green Tea	Catechins, polyphenols	Antioxidant, stress reduction, enhances growth performance	Tilapia, Carp, Catfish
Fenugreek (Trigonella foenum-graecum)	Fenugreek	Saponins, alkaloids	Improves growth, enhances immune system, antioxidant	Tilapia, Carp
Pepper (Piper nigrum)	Black Pepper	Piperine	Improves digestion, antimicrobial, growth promoter	Carp, Tilapia, Salmonids
Clove (Syzygium aromaticum)	Clove	Eugenol	Antimicrobial, antioxidant, enhances digestion	Shrimp, Carp, Salmonids
Neem (Azadirachta indica)	Neem	Azadirachtin, nimbin	Antibacterial, antiparasitic, growth promoter	Tilapia, Catfish, Carp

7. Conclusion

Phytogenics offer a natural and effective alternative to synthetic additives in aquaculture, providing benefits such as improved growth, enhanced immunity, and protection against pathogens. As the industry moves towards more sustainable production systems, the use of plant-based additives will play a key role in addressing the environmental and health challenges associated with intensive fish farming. With continued research and innovation, phytogenics have the potential to shape the future of aquaculture, contributing to a more sustainable and resilient global food system.

References

Abdel-Latif, H.M.R., Khalil, R.H., & Abd-Elhakim, Y.M. (2020). Natural phytobiotics improve the health and performance of tilapia fish. Aquaculture Research, 51(7), 2627-2640.

Ahmadifar, E., Fadaei, R., & Mohammadi, M. (2019). The effects of ginger (Zingiber officinale) on growth performance and health status of aquaculture species: A review. Aquaculture Nutrition, 25(4), 616-627.

Awad, E., & Awaad, A. (2017). Role of medicinal plants on growth performance and immune status in fish. Fish & Shellfish Immunology, 67, 40-54.

Chakraborty, S.B., & Hancz, C. (2011). Application of phytochemicals as immunostimulant, antipathogenic and antistress agents in finfish culture. Reviews in Aquaculture, 3(3), 103-119.

Dawood, M.A.O., & Koshio, S. (2016). Recent advances in the role of probiotics and prebiotics in carp aquaculture: A review. Aquaculture, 454, 243-251.

Diler, I., Yildirim, P., & Metin, S. (2017). The role of garlic (Allium sativum) in aquatic animal health: A review. Aquaculture Nutrition, 23(4), 929-935.

El-Dakar, A.Y., & Goher, M.E. (2015). "Phytogenics as natural growth promoters in aquaculture: Garlic and ginger as case study." International Journal of Fisheries and Aquatic Studies, 2(3), 78-84.

Farahi, A., Kasiri, M., & Sudagar, M. (2014). "The effect of garlic (Allium sativum) supplementation on growth, survival, and stress resistance of rainbow trout (Oncorhynchus mykiss)." Journal of Aquatic Animal Health, 26(2), 63-68.

Hoseinifar, S.H., Sun, Y.Z., & Wang, A. (2018). "Probiotics as means of diseases control in aquaculture." Aquaculture Research, 49(1), 137-150.

Hussain, A., & Jabeen, F. (2015). "The role of phytogenics as natural growth promoters in finfish." Fish Physiology and Biochemistry, 41(5), 1287-1299.

Jalili, R., & Momeni, M. (2019). "Phytogenic feed additives: Enhancing fish growth and immune system using natural products." Aquaculture Nutrition, 25(2), 231-243.

Kalantzi, I., & Houlihan, D.F. (2017). "Phytogenics as natural growth promoters in tilapia: A review." Journal of the World Aquaculture Society, 48(6), 955-970.

Kesarcodi-Watson, A., Kaspar, H., & Lategan, M.J. (2008). "Probiotics in aquaculture: The need, principles and mechanisms of action and screening processes." Aquaculture, 274(1), 1-14.

Lee, J.Y., & Gao, Y. (2012). "Role of dietary phytogenics in modulating fish health and growth performance: An overview." Fish & Shellfish Immunology, 32(3), 409-417.

Lin, Y.H., & Shiu, Y.L. (2019). "Effect of turmeric extract (Curcuma longa) on growth, immune response, and disease resistance in tilapia (Oreochromis niloticus)." Aquaculture Research, 50(2), 516-523.

Nya, E.J., & Austin, B. (2009). "Use of garlic (Allium sativum) to control Aeromonas hydrophila infection in rainbow trout (Oncorhynchus mykiss)." Journal of Fish Diseases, 32(11), 963-970.

Pandey, A.K., & Verma, S. (2018). "Potential of medicinal plants as growth promoters and immunostimulants in aquaculture." Reviews in Aquaculture, 10(3), 576-591.

Puello-Cruz, A.C., & Martínez-Manzano, M.A. (2017). "The potential role of phytobiotics and essential oils in shrimp and fish production." Aquaculture Reports, 6, 57-64.

Reverter, M., Bontemps, N., & Lecchini, D. (2014). "Use of plant extracts in fish aquaculture as an alternative to chemotherapy: Current status and future perspectives." Aquaculture, 433, 50-61.

Sahu, S., & Das, B.K. (2007). "The role of curcumin (Curcuma longa) in modulating fish immune responses." Aquaculture, 272(1), 20-25.

Selim, K.M., & Reda, R.M. (2015). "The use of some medicinal plants as immunostimulants in fish farms." Journal of Aquaculture Research & Development, 6(2), 1-11.

Soleimani, N., & Afsharnasab, M. (2015). "Phytogenics in shrimp and fish nutrition: Benefits and challenges." Aquaculture International, 23(2), 393-409.

Sutili, F.J., & Gatlin, D.M. (2018). "Plant-based compounds as growth promoters and health boosters in fish: A review." Fish & Shellfish Immunology, 74, 1-15.

Wang, W., & Zhou, Z. (2017). "Phytogenics as natural growth enhancers in aquaculture: A review." Journal of Animal Science and Biotechnology, 8(1), 12-22.

Yilmaz, E., & Ergün, S. (2014). "The use of oregano and rosemary as feed additives in aquaculture." Aquaculture International, 22(6), 1469-1478.

6

Feeding for Reproductive Success in Fish: Nutritional Strategies

1. Introduction

Reproduction is one of the most critical processes in the life cycle of fish, directly influencing population sustainability, species survival, and aquaculture productivity. Successful reproduction hinges on several interconnected factors, including environmental conditions (such as water quality, temperature, and photoperiod), genetic predisposition, physiological health, and, most importantly, nutrition. Among these, nutrition has a profound and direct influence on reproductive performance in both wild and cultured fish populations.

In aquaculture, the reproductive success of broodstock—mature fish that produce offspring—is paramount to the production of high-quality fry (young fish) and larvae, which in turn affects the overall yield and profitability of fish farms. Ensuring that broodstock are provided with an optimal diet tailored to their reproductive needs can significantly enhance fecundity (the number of eggs produced), fertilization rates, and the viability of offspring. Proper nutrition not only contributes to better gamete (egg and sperm) quality but also plays a vital role in boosting the health, vitality, and resilience of both broodstock and their progeny.

Nutritional deficiencies or imbalances during the reproductive phase can lead to a variety of problems, such as reduced spawning frequency, poor egg and sperm quality, low hatching rates, and poor larval survival. Conversely, well-nourished broodstock with access to essential nutrients are more likely to produce high-quality eggs and sperm that result in healthier, more viable offspring.

In the wild, fish can access a diverse array of food sources to meet their reproductive needs. However, in aquaculture systems, where broodstock are typically confined to controlled environments, their dietary intake is entirely dependent on formulated feeds. Therefore, developing and implementing scientifically optimized nutritional strategies is critical to ensure the reproductive success of cultured fish species. These strategies must account for the varying nutritional needs of fish during different stages of the reproductive cycle, including pre-spawning, spawning, and post-spawning phases.

This chapter delves into the intricate relationship between nutrition and reproductive success in fish, providing a comprehensive overview of the essential nutrients that support reproductive health, growth, and fertility. It will explore the role of macronutrients such as proteins, lipids, and carbohydrates, as well as micronutrients like vitamins and minerals, in ensuring successful gametogenesis (the formation of gametes), egg quality, and larval development. Moreover, it will discuss the latest feeding strategies and specialized feed formulations designed to optimize broodstock performance in various aquaculture systems.

By understanding the nutritional requirements of fish during reproduction, aquaculture practitioners can make informed decisions about feed management, resulting in enhanced reproductive outcomes, greater larval survival rates, and more efficient production processes. This knowledge is especially important as the global demand for fish continues to rise, making it imperative for the aquaculture industry to adopt sustainable, efficient, and productive practices to meet this growing demand.

In the following sections, we will examine the specific roles of key nutrients, including proteins, fatty acids, vitamins, and minerals, in fish reproduction. We will also explore the challenges faced in broodstock nutrition and highlight future research directions aimed at refining reproductive feeding strategies. Through these insights, this chapter aims to contribute to the understanding of how optimal nutrition can lead to greater reproductive success, improved broodstock health, and sustainable aquaculture production.

2. Nutritional Requirements for Fish Reproduction

Fish reproduction is a biologically demanding process, involving the development of gonads (testes and ovaries), production of high-quality gametes (sperm and eggs), and, in many species, the subsequent care of offspring. To achieve reproductive success, fish must receive appropriate nutrition tailored to meet the physiological demands of each reproductive phase. Nutrient deficiencies or imbalances can lead to suboptimal reproductive performance, affecting egg quality, sperm motility, fertilization rates, and larval survival. This section discusses the critical macronutrients and micronutrients required for optimizing reproductive performance in fish.

2.1. Proteins and Amino Acids

Proteins are essential for tissue growth and development, and their importance is magnified during gametogenesis—the process through which sperm and eggs are formed. Amino acids, which are the building blocks of proteins, play specific roles in reproductive processes:

- **Arginine and Lysine**: These amino acids are essential for sperm motility and fertilization success. Arginine also promotes nitric oxide production, enhancing blood flow to the gonads, which is critical for reproductive health.
- **Methionine and Cysteine**: These sulfur-containing amino acids are crucial for oocyte (egg) maturation and egg membrane development.
- **High-Protein Diets**: Adequate levels of high-quality protein in broodstock diets ensure the development of viable gametes, supporting ovarian and testicular development. Proteins also contribute to the formation of yolk proteins, essential for embryonic development and larval nutrition.

Inadequate protein levels can lead to poor reproductive performance, including reduced egg production, lower fertilization rates, and compromised larval survival.

2.2. Lipids and Fatty Acids

Lipids, particularly essential fatty acids (EFAs), play a vital role in reproductive success. They serve as an energy source and are integral to membrane structure and hormone synthesis:

- **Omega-3 and Omega-6 Fatty Acids**: These polyunsaturated fatty acids (PUFAs) are key to reproductive health. Omega-3 fatty acids, particularly eicosapentaenoic acid (EPA) and docosahexaenoic acid (DHA), are essential for egg quality, embryo development, and larval survival.
- **Energy Reserves**: Lipids act as energy stores for broodstock, especially females, who accumulate fat in their ovaries to fuel egg production. During the spawning season, energy requirements increase, and lipids provide the necessary reserves to sustain reproductive activities.
- **Improved Egg Quality**: Studies have shown that broodstock diets rich in omega-3 fatty acids lead to better egg quality, higher fertilization rates, and improved larval health and survival.

Deficiencies in essential fatty acids can result in poor egg viability, decreased fertilization success, and increased embryonic mortality.

2.3. Carbohydrates

Although carbohydrates do not directly influence reproductive success, they are an important energy source for broodstock during the reproductive process:

- **Energy Supply**: Carbohydrates provide readily metabolizable energy, reducing the need to break down proteins and lipids for energy. This allows these other macronutrients to be utilized for gametogenesis and gonadal development rather than energy production.
- **Prevention of Catabolism**: By supplying adequate carbohydrates, broodstock can avoid the breakdown of muscle tissue and other energy reserves, preserving body condition and reproductive potential.

A balanced inclusion of carbohydrates in the diet ensures that broodstock have sufficient energy during the energy-intensive spawning periods.

2.4. Vitamins

Vitamins are critical co-factors in enzymatic processes, hormone regulation, and immune function, all of which are essential for successful reproduction:

- **Vitamin E**: This antioxidant protects reproductive cells from oxidative damage. It is particularly important for improving sperm motility and egg quality. Vitamin E also enhances reproductive hormone synthesis, promoting successful gametogenesis.
- **Vitamin C**: As a powerful antioxidant, vitamin C helps to prevent oxidative stress in reproductive tissues, improving fertility. It supports collagen synthesis, which is critical for oocyte (egg cell) development and embryo survival.
- **Vitamin A**: Essential for gonadal development and maturation of eggs, vitamin A plays a key role in embryonic development. Deficiencies can result in poor reproductive performance and reduced larval survival.
- **Vitamin D**: Important for calcium regulation, vitamin D supports sperm motility and egg shell formation. Adequate calcium levels are crucial for the hardening of eggs, particularly in oviparous species (egg-laying fish).

2.5. Minerals

Minerals are essential for several reproductive processes, including gamete formation, hormone production, and embryonic development:

- **Calcium and Phosphorus**: These minerals are required for skeletal development and egg shell formation. Calcium is also involved in sperm motility, while phosphorus is important for energy metabolism during reproduction.

- **Zinc**: Zinc supports DNA synthesis and cellular repair, making it crucial for gametogenesis. It also improves sperm motility and is involved in the development of healthy eggs.
- **Selenium**: This mineral acts as an antioxidant, protecting eggs and sperm from oxidative stress, which is particularly important during periods of environmental or physiological stress.
- **Manganese**: Manganese is involved in hormone production and is essential for proper formation of egg shells in oviparous species.

3. Feeding Strategies for Optimal Reproductive Performance

Different stages of the reproductive cycle require specific nutritional strategies to ensure the optimal performance of broodstock. Pre-spawning, spawning, and post-spawning nutritional needs vary significantly, and adjusting feeding regimens accordingly can maximize reproductive success.

3.1. Pre-Spawning Nutrition

Pre-spawning nutrition is critical for preparing broodstock for reproduction. During this stage, the focus is on building up energy reserves and ensuring the development of high-quality gametes:

- **High Protein Content**: Diets should be rich in high-quality protein to support gonadal development and egg production.
- **Essential Fatty Acids**: Omega-3 and omega-6 fatty acids are crucial for the maturation of eggs and sperm. EPA and DHA should be included in the diet to enhance egg quality and larval survival.
- **Antioxidants**: Supplementing with vitamins E and C can reduce oxidative stress, improving the quality of eggs and sperm.

3.2. Spawning Nutrition

During the spawning period, energy demands are higher due to the reproductive activities of the broodstock. Diets should provide sufficient energy and nutrients to maintain reproductive health:

- **Energy-Rich Feeds**: Carbohydrates and lipids should be included in adequate amounts to meet the increased energy requirements.
- **Vitamins and Minerals**: Adequate levels of vitamins and minerals, particularly vitamin E, C, and zinc, are necessary to support hormone production and gamete formation.

3.3. Post-Spawning Nutrition

After spawning, broodstock require nutrition to recover from the energy-intensive reproductive process and to prepare for future reproductive cycles:

- **Protein-Rich Diets**: High-quality protein should be provided to support tissue repair and recovery after spawning.
- **Lipid Replenishment**: Lipid content in the diet should be increased to restore the energy reserves that were depleted during the spawning phase.

4. Nutritional Effects on Egg and Larval Quality

The nutrition provided to broodstock is a fundamental determinant of reproductive outcomes in fish. Adequate nutrition during gametogenesis, especially the period leading up to spawning, directly influences egg quality and larval viability. Nutrient deficiencies or imbalances can compromise reproductive success, leading to smaller or less viable eggs, reduced hatching success, and weak, poorly developing larvae. Optimizing the nutrient profile in broodstock diets is essential to ensure not only high fecundity but also the production of strong, resilient offspring capable of surviving early developmental stages.

4.1. Egg Quality

High-quality eggs are larger, with robust yolk reserves, which are essential for supporting the developing embryo. Egg quality is often evaluated based on key factors such as size, yolk composition, and viability, all of which are influenced by maternal nutrition. Some essential nutrients that directly impact egg quality include:

- **Essential Fatty Acids (DHA and EPA)**: Docosahexaenoic acid (DHA) and eicosapentaenoic acid (EPA), both omega-3 polyunsaturated fatty acids, play a crucial role in the structural integrity and functional capacity of eggs. These fatty acids are incorporated into the egg membrane, ensuring its fluidity and permeability, which are vital for normal embryonic development and metabolic processes. Diets rich in marine sources, such as fish oil and marine algae, supply these key fatty acids, leading to improved egg quality.
- **Yolk Reserves**: The yolk of fish eggs is the primary nutrient reservoir for developing embryos. Lipids, particularly essential fatty acids, are the primary energy source, while proteins in the yolk provide the building blocks for tissue formation during embryonic development. The quality and quantity of yolk reserves are directly influenced by the dietary lipid and protein content consumed by the broodstock.
- **Vitamins (E and C)**: Antioxidants like vitamins E and C are crucial for protecting eggs from oxidative damage during oogenesis (egg development) and after fertilization. Vitamin E enhances egg membrane

stability and improves overall egg quality, while vitamin C supports collagen synthesis, which is important for structural integrity during egg development. These antioxidants help reduce oxidative stress on developing oocytes, increasing the likelihood of producing high-quality eggs that can withstand environmental challenges during early development.

- **Minerals**: Calcium, phosphorus, and zinc are vital for proper egg formation. Calcium and phosphorus are crucial for hardening the egg shell, while zinc supports the DNA synthesis and repair mechanisms that are critical during early embryogenesis.

4.2. Larval Survival

The quality of the egg not only influences embryonic development but also the post-hatching survival of larvae. Nutrient reserves, particularly those stored in the yolk sac during oogenesis, are vital for larval growth during the initial stages of life before they transition to exogenous feeding. Key factors affecting larval survival include:

- **Lipid and Fatty Acid Content**: Lipids and essential fatty acids (DHA and EPA) from the yolk provide energy for the rapidly developing larvae. These nutrients are crucial for organ development, including the brain and nervous system, and for building the structural membranes of cells. Poor maternal nutrition can lead to insufficient lipid reserves, resulting in larvae with reduced energy stores, impaired organ development, and poor growth potential.
- **Immune System Development**: The immune system of fish larvae is immature at hatching, relying heavily on maternal nutrient provisioning for early defense against pathogens. Antioxidants like vitamins E and C, along with minerals like selenium and zinc, contribute to a stronger immune system in larvae, reducing their susceptibility to diseases. Broodstock diets deficient in these nutrients can lead to weakened immune responses, resulting in higher mortality rates among larvae.
- **Growth and Development**: Proper nutritional support during the egg stage ensures that larvae have the metabolic resources to thrive during the critical first days of life. Poor nutrition can result in smaller larvae with reduced growth potential, increasing their vulnerability to environmental stressors and predation.

5. Specialized Broodstock Diets and Formulations

To optimize reproductive performance and improve egg and larval quality, aquaculture has increasingly focused on formulating specialized broodstock diets. These diets are designed to provide an optimal balance of nutrients necessary for successful reproduction, focusing on key components such as high-quality proteins, essential fatty acids, vitamins, and minerals. Common components of specialized broodstock diets include:

- **High-Quality Proteins**: Protein sources like fish meal, soybean meal, and other plant-based proteins ensure an adequate supply of essential amino acids for gametogenesis. The quality and digestibility of these proteins are critical for gonadal development and gamete production.
- **Fish Oil and Marine Algae**: These are rich sources of omega-3 fatty acids (DHA and EPA), which are essential for egg membrane integrity, embryonic development, and larval health. Marine algae, in particular, are gaining attention as a sustainable alternative to fish oil.
- **Prebiotics and Probiotics**: The inclusion of prebiotics and probiotics in broodstock diets helps support gut health, improve nutrient absorption, and enhance immune function, contributing to overall reproductive health and performance.
- **Herbal Supplements (Phytogenics)**: Herbal additives such as garlic, turmeric, and ginger are increasingly being incorporated into broodstock diets to naturally enhance immunity, improve reproductive health, and reduce oxidative stress. These phytogenics offer a sustainable and environmentally friendly alternative to synthetic additives.

6. Challenges and Future Directions

Despite advancements in understanding the nutritional requirements of broodstock and the development of specialized diets, several challenges remain:

- **Species-Specific Diets**: The nutritional needs of broodstock vary significantly among different fish species, and there is no one-size-fits-all solution. Formulating species-specific diets that optimize reproductive success remains a challenge, particularly for less-researched or commercially marginal species.
- **Cost of Specialized Feeds**: High-quality broodstock diets are often expensive due to the inclusion of premium ingredients like fish oil, marine algae, and prebiotics. This can be a barrier for small-scale or subsistence aquaculture operations, limiting their access to these feeds and reducing reproductive outcomes in their stock.
- **Nutritional Interactions**: The complex interplay between different nutrients, such as vitamins, minerals, fatty acids, and proteins, is not fully understood. Further research is needed to optimize nutrient synergy,

ensuring that the right balance of nutrients is delivered in broodstock diets to maximize reproductive performance.

Future Research Directions

To address these challenges, future research should focus on:

- **Optimizing Species-Specific Diets**: Research should aim to better understand the specific nutritional requirements of different fish species during reproduction, allowing for the development of more tailored broodstock diets.
- **Alternative Ingredients**: Exploring alternative, sustainable ingredients such as plant-based proteins and oils, along with microbial and algal sources, can help reduce reliance on marine resources and lower the cost of broodstock feeds.
- **Nutrient Interactions**: Investigating how different nutrients interact during gametogenesis and larval development will improve diet formulations, ensuring that broodstock receive a balanced and effective nutrient profile.
- **Innovative Additives**: Continued exploration of natural additives such as phytogenics, as well as prebiotics and probiotics, can further enhance broodstock health, reproductive success, and sustainability in aquaculture.

7. Conclusion

Feeding for reproductive success in fish is an intricate and vital aspect of aquaculture management, as the quality and health of broodstock directly influence the success of breeding and the viability of the offspring. Achieving optimal reproductive outcomes hinges on providing targeted, nutrient-rich diets that support the unique physiological demands of broodstock at various reproductive stages. These nutritional strategies not only boost reproductive performance but also improve egg quality, enhance larval survival, and contribute to the long-term success and sustainability of aquaculture operations.

Key Nutritional Components for Reproductive Success

- **Proteins and Amino Acids**: Proteins are essential for tissue development and gametogenesis. The role of specific amino acids, such as arginine, lysine, methionine, and cysteine, is critical in enhancing sperm motility and oocyte maturation. High-protein diets are fundamental in ensuring robust ovarian and testicular development, leading to higher fecundity, better fertilization rates, and the production of viable gametes.

- **Essential Fatty Acids (EFAs)**: Omega-3 and omega-6 fatty acids, including DHA and EPA, are critical for egg membrane integrity, embryonic development, and larval survival. These EFAs contribute to the structural integrity of eggs and the development of the nervous system in larvae, thus playing a key role in determining overall reproductive success. Lipids also serve as energy reserves, essential during gametogenesis and larval growth, and their inclusion in broodstock diets ensures better reproductive outcomes.
- **Vitamins and Minerals**: Vitamins, particularly antioxidants like vitamins E and C, and essential minerals such as calcium, phosphorus, zinc, and selenium, are indispensable for reproductive processes. They act as co-factors in hormonal regulation, protect gametes from oxidative stress, and support embryo development. Vitamins also enhance sperm motility, egg quality, and immune function, contributing to overall reproductive health.

Feeding Strategies and Timing

A well-formulated diet alone is insufficient without considering the timing and strategy of feeding during different reproductive phases. Specific feeding strategies, such as pre-spawning, spawning, and post-spawning nutrition, are necessary to cater to the fluctuating physiological demands of broodstock:

- **Pre-Spawning Nutrition**: High-protein and EFA-rich diets prepare broodstock for gametogenesis, promoting the production of high-quality eggs and sperm.
- **Spawning Nutrition**: During the spawning phase, energy demands are heightened, and a balanced provision of carbohydrates, lipids, and essential micronutrients ensures that broodstock can meet these reproductive energy needs without compromising gamete quality.
- **Post-Spawning Nutrition**: After spawning, the focus shifts to replenishing depleted energy reserves and repairing damaged tissues. Protein-rich and lipid-replenishing diets support recovery and prepare the fish for subsequent reproductive cycles.

Impact on Egg and Larval Quality

Proper nutrition extends beyond broodstock health; it has a direct impact on the quality of eggs and the survival of larvae. High-quality eggs, with larger size and better yolk reserves, result in more viable larvae with enhanced survival rates. Optimized broodstock diets improve fecundity, hatching success, and larval growth potential, ultimately enhancing productivity in aquaculture systems. Poor nutrition, on the other hand, can lead to smaller eggs, reduced yolk reserves, and compromised larval health, making offspring more susceptible to diseases, poor growth, and higher mortality rates. By focusing on the nutritional needs of broodstock, aquaculture managers can ensure stronger larvae with better growth performance and resilience.

Broodstock Diets for Sustainable Aquaculture

The formulation of specialized broodstock diets—rich in high-quality proteins, essential fatty acids, antioxidants, and minerals—is an indispensable tool for improving reproductive outcomes. Incorporating sustainable ingredients, such as plant-based proteins and oils, and natural additives like prebiotics, probiotics, and phytogenics, can further enhance the health of broodstock and their offspring while reducing reliance on marine resources. The integration of these nutritionally balanced diets not only increases the reproductive efficiency of broodstock but also supports the sustainability of aquaculture operations by minimizing waste, enhancing fish welfare, and maximizing productivity.

Long-Term Sustainability

As aquaculture continues to expand to meet global demands for fish protein, the importance of broodstock nutrition cannot be overstated. By refining our understanding of species-specific nutritional needs and formulating more efficient, sustainable diets, aquaculture can:

1. **Enhance Productivity**: Nutritional optimization leads to more efficient breeding cycles, larger egg volumes, better fertilization rates, and higher survival rates of larvae, thus improving overall yields.
2. **Promote Fish Welfare**: Adequate nutrition supports the health and vitality of broodstock, reducing stress, enhancing immune function, and ensuring better reproductive performance.
3. **Support Environmental Sustainability**: Sustainable feed formulations that utilize alternative, plant-based ingredients or natural additives help reduce the environmental impact of aquaculture, making it a more eco-friendly industry.

Final Thoughts

In conclusion, feeding for reproductive success in fish is a multifaceted challenge that requires an in-depth understanding of the nutritional needs of broodstock at various stages of reproduction. Properly formulated diets—rich in high-quality proteins, essential fatty acids, vitamins, and minerals—are crucial to enhancing reproductive outcomes, ensuring high egg and larval quality, and supporting the overall sustainability of aquaculture systems. By investing in the nutritional health of broodstock, aquaculture producers can achieve higher productivity, better fish welfare, and contribute to a more sustainable and resilient aquaculture industry for the future.

References

Alava, V. R., & Lim, C. (1983). The quantitative dietary protein requirement of Penaeus monodon juveniles in a controlled environment. Aquaculture, 30(1-4), 53-61.

Bell, J. G., & Sargent, J. R. (2003). Arachidonic acid in aquaculture feeds: current status and future opportunities. Aquaculture, 218(1-4), 491-499.

Bromage, N., & Roberts, R. J. (1995). Broodstock Management and Egg and Larval Quality. Blackwell Science.

Bromage, N., Jones, J., Randall, C., Thrush, M., Davies, B., Springate, J., Duston, J., & Barker, G. (1992). Broodstock management, fecundity, egg quality, and the timing of egg production in the rainbow trout (Oncorhynchus mykiss). Aquaculture, 100(1-3), 141-166.

Brooks, S., Tyler, C. R., & Sumpter, J. P. (1997). Egg quality in fish: What makes a good egg? Reviews in Fish Biology and Fisheries, 7, 387-416.

Cahu, C. L., Zambonino Infante, J. L., & Takeuchi, T. (2003). Nutritional components affecting skeletal development in fish larvae. Aquaculture, 227(1-4), 245-258.

Choubert, G., & Blanc, J. M. (1989). Coloration of rainbow trout (Salmo gairdneri) fed astaxanthin: variations among individual fish. Aquaculture, 81(2), 145-163.

De Silva, S. S., & Anderson, T. A. (1995). Fish Nutrition in Aquaculture. Springer.

Fernández-Palacios, H., Izquierdo, M. S., Robaina, L., Valencia, A., Salhi, M., & Montero, D. (1995). Effect of n-3 HUFA level in broodstock diets on egg quality of gilthead seabream (Sparus aurata L.). Aquaculture, 132(3-4), 325-337.

Hardy, R. W., & Keay, D. (1972). Nutritional and environmental factors affecting the reproductive performance of brook trout (Salvelinus fontinalis). Journal of the Fisheries Board of Canada, 29(7), 1029-1034.

Izquierdo, M. S. (2005). Essential fatty acid requirements in Mediterranean fish species. Cahiers Options Méditerranéennes, 63, 91-102.

Izquierdo, M. S., Fernández-Palacios, H., & Tacon, A. G. J. (2001). Effect of broodstock nutrition on reproductive performance of fish. Aquaculture, 197(1-4), 25-42.

Martínez-Palacios, C. A., Ross, L. G., & Jiménez-Martínez, L. D. (1996). The effects of dietary protein on egg and larval quality in the tilapia (Oreochromis mossambicus). Aquaculture, 145(1-4), 97-108.

Mazorra, C., Bruce, M., Bell, J. G., Davie, A., Alorend, E., Jordan, N., Rees, J., Papanikos, N., Porter, M., & Bromage, N. (2003). Dietary lipid enhancement of broodstock reproductive performance and egg and larval quality in Atlantic halibut (Hippoglossus hippoglossus). Aquaculture, 227(1-4), 21-33.

Morehead, D. T., & Hart, P. R. (2003). Effect of dietary lipid content on egg quality of captive-spawned sand flathead, Platycephalus bassensis. Aquaculture, 214(1-4), 77-90.

Rainuzzo, J. R., Reitan, K. I., & Olsen, Y. (1997). The significance of lipids at early stages of marine fish: a review. Aquaculture, 155(1-4), 103-115.

Rønnestad, I., Thorsen, A., & Finn, R. N. (1998). Fish larval nutrition: a review of recent advances in the roles of amino acids. Aquaculture, 177(1-4), 201-216.

Sargent, J. R., McEvoy, L., & Bell, J. G. (1997). Requirements, presentation and sources of polyunsaturated fatty acids in marine fish larval feeds. Aquaculture, 155(1-4), 117-127.

Teshima, S., Kanazawa, A., & Koshio, S. (1992). Lipid nutrition in fish. Comparative Biochemistry and Physiology Part A: Physiology, 102(4), 778-782.

Tocher, D. R. (2003). Metabolism and functions of lipids and fatty acids in teleost fish. Reviews in Fisheries Science, 11(2), 107-184.

Verakunpiriya, V., Watanabe, T., Ohta, M., & Kiron, V. (1997). Broodstock nutrition and egg quality of red seabream after feeding diets containing eicosapentaenoic or docosahexaenoic acid. Fisheries Science, 63(4), 571-577.

Watanabe, T. (1993). Importance of docosahexaenoic acid in marine larval fish. Journal of the World Aquaculture Society, 24(2), 152-161.

Watanabe, T., Kiron, V., & Satoh, S. (1997). Trace minerals in fish nutrition. Aquaculture, 151(1-4), 185-207.

Watanabe, T., Ohta, M., Kanematsu, M., & Isshiki, T. (1991). Effect of dietary vitamin E on growth, fatty acid composition of muscle and spawning of red seabream broodstock. Nippon Suisan Gakkaishi, 57(2), 345-351.

Zohar, Y., & Mylonas, C. C. (2001). Endocrine manipulations of spawning in cultured fish: from hormones to genes. Aquaculture, 197(1-4), 99-136.

7

Role of Enzymes in Improving Feed Digestibility and Performance

1. Introduction

Enzymes are specialized proteins that act as biological catalysts, facilitating and accelerating biochemical reactions within living organisms. They are essential for various physiological processes, including digestion, metabolism, and energy production. In fish and other aquatic animals, enzymes play a pivotal role in breaking down complex nutrients found in their diets into simpler, absorbable components. This enzymatic action is vital for maximizing nutrient utilization, promoting growth, and ensuring overall health. The inclusion of enzymes in aquaculture feeds has garnered significant attention over recent years, particularly as the industry strives to meet the increasing global demand for fish and seafood products. With aquaculture now accounting for over 50% of the fish consumed worldwide, there is an urgent need for innovative strategies that enhance the sustainability and efficiency of fish farming operations. One such strategy is the supplementation of feeds with exogenous enzymes, which has been shown to improve feed digestibility, promote better growth performance, and mitigate environmental impacts associated with aquaculture.

1.1. The Role of Enzymes in Nutrient Digestion

Fish diets often consist of a combination of protein, carbohydrates, and lipids sourced from both animal and plant materials. However, many of these feed ingredients, especially plant-based sources, contain complex macromolecules that can be difficult for fish to digest. Fish possess a limited capacity for endogenous enzyme production, which can hinder their ability to effectively break down these complex nutrients. As a result, poorly digested feed not only leads to inefficient nutrient absorption but also contributes to increased waste production, further exacerbating water quality issues in aquaculture systems.

The incorporation of exogenous enzymes in aquafeeds addresses these challenges by enhancing the digestibility of feed components. For example, proteases break down proteins into peptides and amino acids, carbohydrases degrade carbohydrates into simple sugars, and lipases facilitate the breakdown of fats into fatty acids and glycerol. By supplementing feeds with these enzymes, aquaculture producers can significantly improve the bioavailability of nutrients, leading to more efficient feed utilization and enhanced growth rates in fish.

1.2. Benefits of Enzyme Supplementation

The benefits of enzyme supplementation in aquaculture extend beyond mere improvements in nutrient digestibility. Enhanced feed utilization translates into better growth performance, characterized by higher feed conversion ratios (FCR), which indicate a more efficient conversion of feed into fish biomass. Additionally, the use of enzymes can help reduce feed costs by allowing the incorporation of lower-cost plant-based ingredients while maintaining optimal growth rates and health. Furthermore, enzyme supplementation has notable environmental benefits. By improving nutrient absorption and minimizing undigested feed waste, enzymes help reduce the nutrient load in aquatic ecosystems, thus lessening the risk of eutrophication and other forms of environmental degradation. This is particularly important in an era where sustainability and environmental stewardship are at the forefront of aquaculture practices.

This chapter will explore the multifaceted role of enzymes in improving feed digestibility and performance in aquaculture. We will delve into the different types of enzymes commonly used in aquafeeds, including proteases, carbohydrases, phytases, and lipases, discussing their specific functions and mechanisms of action. We will also examine the impact of these enzymes on fish growth, health, and overall performance. Additionally, we will review relevant studies that demonstrate the effectiveness of enzyme supplementation across various fish species and feeding regimes. Finally, this chapter will address the challenges associated with enzyme application in aquaculture, as well as future directions for research and development in this area. By understanding the critical role of enzymes in fish nutrition, aquaculture producers can adopt more effective feeding strategies that optimize feed utilization, enhance fish welfare, and contribute to the sustainability of fish farming operations.

2. Importance of Feed Digestibility in Aquaculture

Feed constitutes one of the largest operational costs in aquaculture, typically representing 50-60% of total production expenses. As aquaculture continues to expand to meet the growing global demand for seafood, the economic pressure to optimize feed utilization becomes paramount. Maximizing feed digestibility is crucial for several reasons, encompassing both economic and environmental considerations.

2.1. Economic Significance

The financial implications of feed digestibility in aquaculture are substantial. When feed is poorly digested, a significant portion of the nutrients remains unutilized, leading to wastage. This inefficiency not only increases feed costs but can also reduce overall profitability. Enhancing feed digestibility through the addition of enzymes enables fish to absorb more nutrients from their diets, resulting in:

- **Improved Growth Rates**: Higher digestibility facilitates better nutrient absorption, leading to accelerated growth in fish. Faster growth rates translate into shorter production cycles and increased marketable biomass.
- **Better Feed Conversion Ratios (FCR)**: FCR is a critical metric in aquaculture that measures the efficiency of feed utilization. A lower FCR indicates that less feed is required to produce a unit of fish weight. Improved feed digestibility contributes to lower FCR, making production more cost-effective.
- **Enhanced Feed Efficiency**: By maximizing the nutritional value derived from feeds, aquaculture producers can achieve higher levels of efficiency in fish farming operations. This efficiency is essential for maintaining competitiveness in an increasingly saturated market.

2.2. Environmental Considerations

Beyond economic implications, poorly digested feeds can have detrimental effects on the environment. When excess nutrients from uneaten feed or metabolic waste are released into the aquatic environment, they can lead to nutrient loading, which contributes to problems such as:

- **Eutrophication**: Eutrophication occurs when water bodies become enriched with excess nutrients, leading to algal blooms, oxygen depletion, and adverse effects on aquatic ecosystems. These blooms can disrupt fish populations, harm other aquatic organisms, and negatively impact water quality.
- **Water Quality Deterioration**: Increased nutrient levels can lead to a decline in water quality, affecting the health and welfare of farmed fish. Poor water quality can also necessitate additional management practices and costs, such as water treatment and increased monitoring.

Improving feed digestibility not only enhances nutrient absorption by fish but also mitigates the environmental impact of aquaculture operations. By utilizing feed more effectively, producers can reduce waste outputs and the subsequent nutrient load on surrounding ecosystems. This approach aligns with the broader goals of sustainability and responsible aquaculture practices, which are increasingly demanded by consumers and regulatory bodies alike.

3. Types of Enzymes Used in Aquafeeds

The inclusion of enzymes in aquaculture feeds has been shown to significantly enhance feed digestibility and performance. Various types of enzymes target specific macromolecules in feed ingredients, ensuring efficient nutrient breakdown. This section outlines the primary enzymes utilized in aquafeeds and their respective roles.

3.1. Proteases

Proteases are enzymes that catalyze the hydrolysis of proteins into smaller peptides and amino acids. This is especially important in aquaculture, where high-quality protein sources are critical for fish growth and development. The addition of exogenous proteases to fish diets enhances protein digestibility, making it more accessible for absorption.

- **Functionality**: Proteases improve the breakdown of complex proteins, particularly those found in plant-based ingredients that may be less digestible than traditional fish meal. By increasing the availability of essential amino acids, proteases contribute to improved growth rates and overall fish health.
- **Impact on Plant-Based Diets**: In aquaculture, there is a growing trend toward incorporating plant-based protein sources to reduce reliance on fish meal. Exogenous proteases help maximize the nutritional value of these ingredients, enabling effective protein utilization.

3.2. Carbohydrases

Carbohydrases, including enzymes such as amylase, xylanase, and glucanase, specifically target carbohydrates in feed ingredients. Fish have limited endogenous carbohydrases, making it challenging to digest certain complex carbohydrates, particularly non-starch polysaccharides (NSPs) commonly found in plant materials.

- **Amylase**: This enzyme breaks down starches into simpler sugars, enhancing energy availability from carbohydrate-rich feeds.
- **Xylanase and Glucanase**: These enzymes act on xylans and glucans, respectively, which are structural components of plant cell walls. By degrading these polysaccharides, carbohydrases improve the digestibility and energy utilization of plant-based feed ingredients.

3.3. Phytases

Phytases are specialized enzymes that hydrolyze phytic acid, a form of phosphorus that is indigestible to fish. In plant-based diets, a significant portion of phosphorus is bound in phytate form, rendering it unavailable for absorption.

- **Liberating Phosphorus**: Phytase enzymes liberate bound phosphorus, making it accessible for uptake. This not only enhances phosphorus utilization for vital functions such as bone development and metabolic processes but also reduces the need for inorganic phosphorus supplements, lowering feed costs.
- **Environmental Benefits**: The use of phytase also decreases the phosphorus discharge into the environment, addressing one of the critical challenges in aquaculture and contributing to more sustainable practices.

3.4. Lipases

Lipases are enzymes responsible for breaking down dietary fats into free fatty acids and glycerol. In fish nutrition, lipases enhance the digestibility of lipids, which are essential for energy, cellular membrane integrity, and hormone production.

- **Maximizing Fat Utilization**: The inclusion of lipases in aquafeeds is particularly beneficial in diets rich in plant oils or alternative lipid sources. By improving fat digestibility, lipases facilitate energy absorption and promote optimal growth.

3.5. Fibrolytic Enzymes

Fibrolytic enzymes, such as cellulase and hemicellulase, are crucial for breaking down fibrous components of plant materials, particularly cellulose and hemicellulose.

- **Increasing Digestibility**: These enzymes target the indigestible fibrous components of plant feeds, increasing nutrient absorption and reducing waste. By enhancing the digestibility of these fibrous materials, fibrolytic enzymes allow fish to utilize a broader range of feed ingredients more effectively.
- **Reducing Feed Waste**: Improved digestibility of fibrous components contributes to lower waste outputs, promoting a cleaner and more sustainable aquaculture environment.

4. Mechanisms of Action of Enzymes

Enzymes are vital biological catalysts that facilitate biochemical reactions in living organisms, including fish. In aquaculture, they play a crucial role in enhancing the digestibility and bioavailability of nutrients in feed. This section delves into the specific mechanisms by which various enzymes operate, highlighting their substrate specificity and their contributions to improving nutrient absorption and overall fish performance.

4.1. Enzyme Catalysis

Enzymes function by catalyzing the hydrolysis of specific chemical bonds within complex feed ingredients. This hydrolysis involves the addition of water to break down larger macromolecules (such as proteins, carbohydrates, and lipids) into smaller, absorbable units (e.g., amino acids, simple sugars, and fatty acids). The catalytic process occurs in several stages:

1. **Substrate Binding**: Each enzyme has a unique active site that binds to its specific substrate. The specificity of enzymes is due to their unique three-dimensional structures, which complement the shape of their target substrates.
2. **Formation of the Enzyme-Substrate Complex**: Upon binding, an enzyme-substrate complex is formed, stabilizing the substrate and positioning it for the catalytic reaction. This complex is essential for the enzyme to exert its catalytic effect.
3. **Catalysis**: The enzyme facilitates the breaking of specific bonds in the substrate, leading to the hydrolysis of macromolecules. This process often involves the formation of transient intermediate structures and may require co-factors or co-enzymes for optimal activity.
4. **Release of Products**: After the reaction, the enzyme releases the products (such as amino acids, sugars, or fatty acids) while remaining unchanged and ready to catalyze further reactions with new substrate molecules.

4.2. Substrate Specificity of Enzymes

The substrate specificity of enzymes is critical in aquaculture, particularly when considering the diverse feed ingredients used in fish diets. Different enzymes target distinct nutrient classes, ensuring that complex macromolecules are efficiently broken down into smaller, more absorbable forms. Here are some key enzymes and their specific mechanisms of action:

4.2.1. Proteases

Mechanism of Action: Proteases specifically target peptide bonds in proteins, hydrolyzing them into smaller peptides and free amino acids. The primary steps involved are:

- **Hydrolysis of Peptide Bonds**: Proteases cleave peptide bonds between amino acids in protein molecules, which is crucial for improving protein digestibility. For instance, exogenous proteases added to fish diets can enhance the digestibility of plant proteins that may otherwise be poorly utilized.
- **Absorption of Amino Acids**: The resulting peptides and amino acids are readily absorbed through the intestinal wall of fish, contributing to growth, tissue repair, and overall health.

4.2.2. Carbohydrases

Mechanism of Action: Carbohydrases, such as amylase, xylanase, and glucanase, target various carbohydrates present in plant-based feeds.

- **Breakdown of Starch**: Amylase catalyzes the hydrolysis of starch molecules into simpler sugars like maltose and glucose, enhancing the availability of energy.
- **Degradation of Non-Starch Polysaccharides (NSPs)**: Xylanase and glucanase act on xylans and glucans, respectively, breaking down structural polysaccharides in plant cell walls, thereby improving nutrient accessibility and energy utilization.

4.2.3. Phytases

Mechanism of Action: Phytases are crucial for hydrolyzing phytic acid (myo-inositol hexakisphosphate), which is the primary storage form of phosphorus in many plant-based feeds.

- **Release of Free Phosphorus**: Phytase cleaves the phosphate groups from phytate, converting it into more bioavailable forms. This process enhances phosphorus absorption in fish, crucial for growth, bone development, and metabolic functions.
- **Reduction of Phytate-Induced Nutritional Constraints**: The inclusion of phytase in feeds not only improves phosphorus bioavailability but also reduces the reliance on inorganic phosphorus supplements, leading to cost savings and reduced environmental impact.

4.2.4. Lipases

Mechanism of Action: Lipases catalyze the hydrolysis of lipids (fats) into free fatty acids and glycerol.

- **Fat Breakdown**: Lipases act on triglycerides and phospholipids in the diet, enhancing the digestibility of dietary fats. This is particularly important in feeds rich in plant oils or alternative lipid sources.
- **Increased Energy Utilization**: The release of free fatty acids from lipids allows for efficient energy absorption, supporting fish growth and metabolic processes.

4.2.5. Fibrolytic Enzymes

Mechanism of Action: Fibrolytic enzymes, such as cellulase and hemicellulase, specifically target fibrous components in plant materials.

- **Degradation of Cellulose and Hemicellulose**: These enzymes catalyze the breakdown of cellulose and hemicellulose, which are typically indigestible for fish. By increasing the digestibility of fibrous materials, fibrolytic enzymes enable better nutrient absorption and reduce feed waste.

4.3. Enhancement of Nutrient Bioavailability

By improving the breakdown of complex feed components, enzymes significantly enhance the bioavailability of nutrients. This is especially important in aquaculture, where fish diets often rely on a mix of animal and plant-based ingredients. The key mechanisms through which enzymes enhance nutrient bioavailability include:

- **Increased Surface Area for Absorption**: The hydrolysis of macromolecules results in smaller particles with a greater surface area, facilitating absorption through the intestinal epithelium.
- **Reduction of Anti-Nutritional Factors**: Many plant ingredients contain anti-nutritional factors that inhibit nutrient absorption (e.g., tannins, lectins). Enzymes can help degrade these compounds, improving the overall digestibility of the feed.
- **Synergistic Effects with Endogenous Enzymes**: The activity of exogenous enzymes can complement the fish's endogenous digestive enzymes, particularly in diets rich in plant-based ingredients that require additional enzymatic support for effective digestion.

5. Benefits of Enzyme Supplementation in Aquaculture Feeds

Enzyme supplementation in aquaculture feeds presents numerous benefits that enhance fish growth, feed efficiency, and overall sustainability of aquaculture practices. The strategic incorporation of enzymes addresses key challenges associated with feed utilization, particularly in diets that heavily rely on plant-based ingredients. This section outlines the various advantages of enzyme supplementation in aquaculture feeds.

5.1. Improved Growth Performance

The incorporation of enzymes into aquafeeds has been widely documented to significantly enhance growth performance across different fish species. Key factors contributing to this improved growth include:

- **Enhanced Nutrient Digestibility**: Enzymes facilitate the breakdown of complex macromolecules into simpler, absorbable forms, allowing fish to effectively utilize the proteins, fats, and carbohydrates present in their diets. For instance, proteases enhance protein digestibility, making amino acids more readily available for growth and tissue repair.
- **Biomass Accumulation**: By maximizing the absorption of essential nutrients, enzymes contribute to increased biomass accumulation. Research has shown that fish fed enzyme-supplemented diets exhibit improved growth rates compared to those on non-supplemented feeds, leading to better production outcomes for aquaculture operations.
- **Optimal Utilization of Plant-Based Ingredients**: In modern aquaculture, there is a growing reliance on plant-based feeds due to the declining availability of marine resources. Enzyme supplementation, particularly with proteases and carbohydrases, enhances the digestibility of plant proteins and carbohydrates, allowing fish to thrive on diets that might otherwise be suboptimal without such enzymatic assistance.

5.2. Enhanced Feed Conversion Ratio (FCR)

The feed conversion ratio (FCR) is a critical metric in aquaculture that indicates the efficiency of feed utilization in converting feed into fish biomass. Enzyme supplementation leads to:

- **Lower FCR Values**: Enzymes improve nutrient absorption and utilization, resulting in lower FCR values. A lower FCR means that fish require less feed to gain weight, which is economically advantageous for producers.
- **Improved Feed Efficiency**: By enhancing nutrient bioavailability, enzymes optimize feed efficiency, enabling fish to achieve better growth with the same or reduced feed inputs. This efficiency is particularly important in the context of rising feed costs and the need for sustainable aquaculture practices.

5.3. Reduced Feed Costs

Enzyme supplementation in aquafeeds can lead to significant cost savings for aquaculture operations:

- **Inclusion of Lower-Cost Ingredients**: Enzymes allow for the effective utilization of lower-cost, plant-based ingredients that may otherwise be less digestible. For example, protease supplementation enhances the digestibility of plant proteins, reducing the need for more expensive animal-based protein sources such as fish meal.
- **Reduced Phosphorus Supplements**: Phytase enzymes can liberate phosphorus from phytate, enabling aquafeeds to incorporate more plant-based ingredients while minimizing the need for inorganic phosphorus supplements. This not only lowers feed costs but also reduces the environmental impact associated with excess phosphorus discharge.

5.4. Environmental Benefits

Enzyme supplementation has notable environmental benefits that contribute to the sustainability of aquaculture:

- **Mitigation of Nutrient Pollution**: Undigested feed ingredients excreted by fish can contribute to nutrient pollution in aquatic ecosystems, leading to problems such as eutrophication. By enhancing nutrient digestibility, enzymes reduce the amount of waste excreted into the environment.
- **Reduced Eutrophication Risks**: The use of phytase reduces the amount of undigested phosphorus discharged into water bodies, thereby mitigating the risk of eutrophication and water quality degradation. This is crucial for maintaining healthy aquatic ecosystems and ensuring the sustainability of aquaculture practices.

5.5. Improved Health and Immunity

In addition to enhancing feed digestibility, certain enzymes can have positive effects on fish health and immunity:

- **Reduction of Anti-Nutritional Factors**: Many plant-based feed ingredients contain anti-nutritional factors, such as protease inhibitors and lectins, which can hinder digestion and nutrient absorption. Enzymes that target these compounds can help reduce their negative effects, promoting better gut health.
- **Support for Immune Function**: Improved digestion and nutrient absorption support the overall health of fish, including their immune system. Fish receiving enzyme-supplemented diets may exhibit enhanced resistance to diseases and infections, contributing to better survival rates and growth performance.

6. Case Studies of Enzyme Use in Aquaculture

Enzymes have become a pivotal component in modern aquaculture, providing various benefits across different species. Here, we explore several notable case studies highlighting the effects of enzyme supplementation in tilapia, salmonids, and shrimp, showcasing the positive impacts on growth performance, feed efficiency, and environmental sustainability.

6.1. Tilapia

Tilapia, a widely cultured freshwater fish species, has shown significant improvements in performance with enzyme supplementation:

- **Study Overview**: In various trials, enzymes such as proteases, phytases, and carbohydrases have been integrated into tilapia diets, particularly those formulated with high levels of plant ingredients.
- **Findings**
    - **Growth Performance**: Research indicates that enzyme-supplemented diets enhance tilapia growth rates. For example, a study found that tilapia fed with diets supplemented with exogenous proteases exhibited higher weight gain compared to those on unsupplemented diets.
    - **Feed Efficiency**: Enzyme supplementation has also improved feed conversion ratios (FCR) in tilapia, allowing for more efficient utilization of nutrients.
    - **Phosphorus Retention**: The inclusion of phytase enabled fish to utilize phosphorus more effectively from plant sources. This led to a notable reduction in the need for inorganic phosphorus supplements while ensuring optimal growth and bone development in tilapia, minimizing the environmental impact of phosphorus discharge.
- **Conclusion**: These findings underscore the potential of enzymes to enhance nutrient utilization and reduce reliance on expensive mineral supplements in tilapia aquaculture.

6.2. Salmonids

Salmonids, including species such as salmon and trout, have been the focus of enzyme research aimed at improving the digestibility of alternative protein sources:

- **Study Overview**: Enzyme supplementation, particularly with proteases and phytases, has been applied to diets containing significant proportions of soybean meal and other plant proteins.

- **Findings**
  - **Growth Rates**: Studies indicate that salmonids receiving enzyme-supplemented diets demonstrate improved growth rates, often achieving comparable or superior performance to those fed traditional fishmeal-based diets.
  - **Feed Conversion Ratio (FCR)**: Enzyme-supplemented diets resulted in lower FCR values, highlighting improved feed efficiency. For example, one study reported a 10-15% reduction in FCR for salmon fed diets with added enzymes compared to controls.
  - **Phosphorus Pollution Reduction**: The application of phytase not only improved phosphorus digestibility but also significantly reduced phosphorus excretion into the environment, addressing concerns related to nutrient pollution in recirculating aquaculture systems.
- **Conclusion**: The integration of enzymes in salmonid feeds exemplifies a successful strategy to optimize feed utilization, promote growth, and mitigate environmental impacts associated with aquaculture practices.

6.3. Shrimp

In shrimp aquaculture, enzyme supplementation has shown promising results in improving nutrient digestibility and overall performance:

- **Study Overview**: Research has focused on the effects of protease and carbohydrase supplementation in shrimp diets, particularly those high in plant proteins.
- **Findings**:
  - **Growth Performance**: Enzyme-supplemented diets have been linked to enhanced growth rates in shrimp, with studies showing significant improvements in weight gain compared to control groups.
  - **Survival Rates**: Enhanced digestibility from enzyme inclusion has also led to higher survival rates in juvenile shrimp, as the improved nutrient absorption contributes to better health and resilience against diseases.
  - **Feed Conversion Efficiency**: Enzyme supplementation has been associated with increased feed conversion efficiency, allowing for better utilization of plant protein sources, which is crucial given the rising costs of fishmeal and other animal protein sources.
- **Conclusion**: The application of enzymes in shrimp feeds illustrates their potential in optimizing crustacean nutrition and improving aquaculture sustainability.

7. Challenges and Future Directions

Despite the advantages observed with enzyme supplementation in aquaculture, several challenges need to be addressed:

- **Heat Stability**: A significant concern is the stability of enzymes during feed manufacturing processes, such as extrusion and pelleting, which often involve high temperatures. Many commercial enzymes lose activity under these conditions, necessitating the development of heat-stable enzyme formulations that can withstand processing.
- **Species-Specific Responses**: The effectiveness of enzyme supplementation varies between species, indicating the need for tailored enzyme applications based on specific fish or shrimp species. More research is required to determine optimal enzyme types, dosages, and combinations for different aquaculture species.
- **Cost Considerations**: While the benefits of enzyme supplementation can outweigh the costs, particularly in large-scale operations, the economic feasibility remains a concern for small-scale farmers. Developing cost-effective enzyme products will be crucial for broader adoption.

Future research should focus on

- **Heat-Stable and Cost-Effective Enzyme Formulations**: Investigating new enzyme formulations that maintain activity post-processing while remaining economically viable for aquaculture producers.
- **Synergistic Effects of Enzymes**: Exploring the combined effects of various enzymes to enhance overall feed performance and address specific nutritional challenges in aquaculture feeds.
- **Novel Enzymes Targeting Anti-Nutritional Factors**: Researching the development of novel enzymes that specifically target anti-nutritional factors in alternative feed ingredients, further improving feed digestibility and nutrient absorption.

8. Conclusion

Enzymes are instrumental in enhancing feed digestibility and performance in aquaculture. By facilitating the breakdown of complex nutrients into simpler forms, enzymes improve nutrient bioavailability, bolster growth rates, and reduce feed costs and environmental impact. The positive outcomes from enzyme supplementation in species such as tilapia, salmonids, and shrimp exemplify the critical role enzymes play in optimizing aquaculture feed formulations, particularly as the industry seeks to become more sustainable and efficient. As research continues to evolve, the incorporation of enzymes in aquafeeds is poised to make significant contributions to the future of aquaculture.

References

AOAC. (2016). Official Methods of Analysis. Association of Official Analytical Chemists.

Aksu, M., & Yilmaz, E. (2020). Effects of enzyme supplementation on growth performance and feed utilization in Nile tilapia (Oreochromis niloticus) fed high soybean meal diets. Aquaculture Reports, 18, 100466.

Baird, D. J., & Sibley, P. (2018). The role of enzymes in aquaculture feed formulation. Aquaculture Nutrition, 24(4), 1013-1024.

Bøgwald, J., & Krogdahl, Å. (2016). The role of dietary enzymes in the nutritional strategy for sustainable aquaculture. Aquaculture, 460, 158-162.

Cera, M. L., et al. (2018). Effects of dietary enzymes on growth performance and nutrient utilization in juvenile white shrimp (Litopenaeus vannamei). Aquaculture, 495, 738-743.

de Souza, C. E. A., et al. (2019). Effect of protease and phytase supplementation on nutrient digestibility and growth performance of tilapia (Oreochromis niloticus) fed diets containing different levels of soybean meal. Aquaculture International, 27(3), 885-896.

Denison, R. H., et al. (2020). The effects of enzyme supplementation on nutrient digestibility and growth performance of hybrid striped bass (Morone chrysops × M. saxatilis) fed plant-based diets. Aquaculture, 523, 735168.

Dias, J., et al. (2017). Dietary supplementation of enzymes improves the growth performance of juvenile tambaqui (Colossoma macropomum) fed diets based on cassava meal. Aquaculture Research, 48(3), 1010-1019.

Enes, P., et al. (2015). The role of enzyme supplementation in improving the nutritional quality of aquaculture feeds. Aquaculture Nutrition, 21(1), 78-90. https://doi.org/10.1111/anu.12101

Hossain, M. M., et al. (2017). Enzyme supplementation in aquaculture feeds: A review. Journal of Aquaculture Research and Development, 8(5), 1-7.

Huang, W., et al. (2021). Effect of dietary enzymes on growth performance and feed utilization in grass carp (Ctenopharyngodon idella). Aquaculture Research, 52(1), 1-10.

Jagannath, S., et al. (2018). Role of enzymes in aquafeed formulation: A review. Aquaculture Reports, 12, 114-121.

Krogdahl, Å., et al. (2013). The role of enzymes in improving the digestibility of plant-based feeds for fish. Aquaculture Nutrition, 19(4), 314-329.

Li, Y., et al. (2019). Effect of enzyme supplementation on the growth performance of fish: A meta-analysis. Fish Physiology and Biochemistry, 45(2), 399-410.

Lunger, A. N., et al. (2019). The impact of dietary enzyme supplementation on the nutrient digestibility and growth performance of juvenile yellow perch (Perca flavescens). Aquaculture Nutrition, 25(1), 245-257.

NRC. (2011). Nutrient Requirements of Fish and Shrimp. National Research Council, The National Academies Press.

Pacheco, A. R., et al. (2020). Effect of enzyme supplementation on growth and nutrient utilization in juvenile Nile tilapia (Oreochromis niloticus) fed diets with varying levels of dietary protein. Aquaculture Nutrition, 26(3), 1020-1031.

Perez, A., et al. (2016). The effect of enzyme supplementation on the growth performance of African catfish (Clarias gariepinus) fed diets containing different levels of cassava. Aquaculture Research, 47(2), 342-352. https://doi.org/10.1111/are.12506

Pohl, A., et al. (2018). Enzymatic approaches for improving feed digestibility in aquaculture: A review. Aquaculture, 495, 166-179.

Rojas, R., et al. (2019). Role of enzyme supplementation in aquaculture feeds: Impact on growth performance and feed conversion ratio in juvenile fish. Journal of Animal Physiology and Animal Nutrition, 103(4), 1282-1295.

Sang, H. M., et al. (2018). Effects of enzyme supplementation on the growth performance of common carp (Cyprinus carpio) fed diets based on different levels of soybean meal. Aquaculture Research, 49(1), 201-210.

Wang, Y., et al. (2020). Effects of dietary enzyme supplementation on growth performance and nutrient digestibility in Asian seabass (Lates calcarifer). Aquaculture Research, 51(5), 1861-1871.

Wu, Y., et al. (2019). Effects of enzyme supplementation on growth performance, feed utilization, and digestive enzyme activity in juvenile grass carp (Ctenopharyngodon idella). Aquaculture Nutrition, 25(1), 207-216.

Xu, H., et al. (2021). Effects of dietary enzymes on growth performance and nutrient digestibility of juvenile red drum (Sciaenops ocellatus) fed diets with different carbohydrate sources. Aquaculture Reports, 20, 100730.

Zhan, F., et al. (2020). Effects of enzyme supplementation on the growth performance and nutrient utilization in juvenile Nile tilapia (Oreochromis niloticus) fed high fishmeal diets. Aquaculture Research, 51(3), 1219-1230.

8

Nutritional Strategies to Improve Larval and Juvenile Fish Survival

1. Introduction

The early life stages of fish, particularly the larval and juvenile phases, are critical periods in aquaculture characterized by rapid physiological changes, high metabolic demands, and heightened vulnerability to external stressors. Mortality rates during these stages can be alarmingly high, with reports indicating that up to 70% or more of fish larvae and juveniles may not survive to reach adulthood. These losses are typically caused by a combination of nutritional deficiencies, environmental stress, suboptimal husbandry practices, and disease outbreaks. Addressing these challenges is paramount for improving survival rates and ensuring the success of aquaculture systems. Nutritional management during the larval and juvenile stages plays a pivotal role in determining the overall health, growth performance, and survival of fish. Fish larvae and juveniles have distinct dietary requirements compared to adults due to their rapid growth, underdeveloped digestive systems, and the need for specific nutrients to support organ development, skeletal formation, and immune system maturation. A comprehensive understanding of these nutritional needs is essential for developing feeding strategies that optimize fish health and maximize growth potential.

1.1. Importance of Early Nutrition in Fish Development

The nutritional status of larval and juvenile fish significantly influences their physiological development and long-term survival. During the early stages of life, fish undergo rapid cell proliferation and differentiation, especially in tissues such as the muscle, digestive organs, and immune system. Proper nutrition is essential to support these processes, ensuring that fish achieve optimal growth rates and develop strong resistance to environmental stressors and diseases.

One of the primary factors affecting larval and juvenile fish survival is their ability to efficiently utilize available nutrients for growth and energy. This is largely determined by the functionality of their digestive system, which is underdeveloped in larvae and gradually matures as fish transition into the juvenile stage. As a result, early-stage fish are particularly sensitive to both the quantity and quality of nutrients they receive, making the choice of feed, feeding frequency, and nutrient composition critical to their survival.

1.2. Challenges in Larval and Juvenile Fish Nutrition

Fish larvae are initially dependent on endogenous energy reserves, such as yolk sac nutrients, which provide them with essential proteins, lipids, and micronutrients necessary for their initial growth and development. However, as these reserves deplete, larvae must transition to exogenous feeding, typically within a few days post-hatch. This transition period is particularly challenging because larvae often exhibit poor feeding behavior and have limited capacity to digest complex feed components. Consequently, the early days of exogenous feeding represent a "nutritional bottleneck," during which survival rates can be significantly impacted.

Some of the key challenges in larval and juvenile fish nutrition include:

- **Inadequate digestive capacity**: Fish larvae possess immature digestive organs and enzymes, which limits their ability to break down complex nutrients. As a result, easily digestible and nutrient-rich feeds are required to meet their nutritional needs.
- **Dietary imbalances**: Improperly formulated feeds can lead to nutritional deficiencies or imbalances that hinder growth and increase susceptibility to diseases. Specific nutrients such as essential fatty acids (EFAs), amino acids, vitamins, and minerals must be provided in the correct proportions to ensure proper development.
- **Variability in feed availability**: Live feeds, which are often essential for early larval stages, may not always be available in sufficient quantity or quality. The nutritional composition of live feeds can also vary depending on the species and the conditions under which they are cultured.
- **Stress factors**: Environmental factors such as water quality, temperature fluctuations, and crowding can exacerbate nutritional deficiencies and reduce the overall resilience of fish larvae and juveniles. Stressful conditions often result in reduced feed intake and poor nutrient absorption.

1.3. Nutritional Approaches to Improve Survival

Effective nutritional strategies aim to address these challenges by providing fish larvae and juveniles with the necessary nutrients to support growth, immunity, and resilience against environmental stressors. These strategies involve the use of high-quality live feeds, nutritionally balanced formulated diets, and supplements that enhance digestive efficiency and nutrient bioavailability.

- **Live feeds**: Microalgae, rotifers, copepods, and Artemia are commonly used as live feeds in the early larval stages. These feeds are highly digestible and provide essential nutrients such as proteins, lipids, and EFAs that are critical for larval development. Live feeds are often enriched with additional nutrients to ensure that they meet the specific needs of fish larvae.
- **Formulated diets**: As larvae transition to the juvenile stage, formulated diets play an increasingly important role in providing complete nutrition. Advances in feed technology have led to the development of microparticulate diets that are specifically designed for larvae and juveniles. These diets are often supplemented with enzymes, probiotics, and other additives to enhance nutrient absorption and promote gut health.
- **Nutritional supplements**: Probiotics, prebiotics, digestive enzymes, antioxidants, and immunostimulants are increasingly being incorporated into larval and juvenile diets to boost survival rates. These supplements help to improve digestive efficiency, enhance immune responses, and reduce the risk of disease outbreaks during these vulnerable life stages.

1.4. Significance of Nutritional Strategies for Sustainable Aquaculture

Sustainability is a key concern in modern aquaculture, with an increasing emphasis on reducing feed costs, minimizing environmental impact, and improving resource efficiency. Nutritional strategies that enhance the survival and growth of larval and juvenile fish contribute to more efficient use of feed resources, thereby reducing overall production costs. Furthermore, by promoting healthy growth and reducing disease outbreaks, these strategies help to minimize the need for chemical treatments and antibiotics, supporting more sustainable and environmentally friendly aquaculture practices.

The development of cost-effective and nutritionally balanced feeds that incorporate alternative protein sources, such as plant-based ingredients, insect meals, and algal biomass, is also critical for reducing the reliance on traditional fishmeal and fish oil, which are finite and environmentally taxing resources.

This chapter explores various nutritional strategies designed to improve the survival, growth, and overall health of larval and juvenile fish. It provides an in-depth analysis of the nutritional requirements of fish during these early life stages, focusing on proteins, lipids, carbohydrates, vitamins, and minerals.

Additionally, the chapter discusses the role of live feeds, formulated diets, and nutritional supplements, as well as emerging trends in feed technology and nutritional science. Special attention is given to innovative approaches such as the use of probiotics, prebiotics, and alternative feed ingredients, which are shaping the future of larval and juvenile nutrition in aquaculture.

By understanding and applying these strategies, aquaculturists can significantly reduce mortality rates, enhance growth performance, and contribute to the overall sustainability and profitability of fish farming operations.

2. Nutritional Requirements of Larval and Juvenile Fish

The nutritional needs of fish larvae and juveniles are unique due to their rapid growth rates and high metabolic demands. These early life stages are characterized by specific requirements for macronutrients (proteins, lipids, carbohydrates) and micronutrients (vitamins, minerals) that are essential for supporting growth, organ development, and immune function. Properly meeting these requirements is crucial to maximizing survival, growth, and overall health, especially in an aquaculture setting where nutritional management can directly influence the success of a farming operation.

2.1. Protein and Amino Acids

Protein is a critical component of larval and juvenile fish diets, constituting the primary macronutrient required for growth, muscle development, and tissue repair. Fish larvae have exceptionally high protein needs, with optimal dietary protein levels often ranging between 45% to 60% of their total intake, depending on the species. This high requirement is driven by the larvae's rapid growth and elevated metabolic rate, both of which necessitate a steady supply of amino acids—the building blocks of proteins.

Essential Amino Acids (EAA) such as methionine, lysine, threonine, arginine, and histidine are especially important. Unlike non-essential amino acids, fish cannot synthesize EAAs de novo and must acquire them from their diet. Deficiencies in any of these essential amino acids can lead to reduced growth performance, impaired immune function, and increased vulnerability to stress and disease. For example, methionine is involved in cellular processes such as methylation and antioxidation, while lysine is important for protein synthesis and immune responses. In larval diets, live feeds such as rotifers and Artemia are highly valued due to their complete amino acid profiles, which naturally meet the nutritional needs of developing larvae. These live feeds contain all the essential amino acids, facilitating optimal growth during the early life stages when the larvae's digestive system is still maturing and not yet efficient at breaking down more complex formulated feeds.

As fish transition to the juvenile stage, their digestive capacity increases, allowing for the inclusion of a wider variety of protein sources in formulated feeds. Common protein sources in juvenile diets include fishmeal, a highly digestible ingredient that provides an excellent balance of amino acids, and soybean meal, which is often used as a plant-based alternative. However, the anti-nutritional factors present in soybean meal, such as trypsin inhibitors, may reduce protein digestibility and require processing or enzyme supplementation to enhance its use in juvenile fish diets. Balancing protein quality and quantity is essential to avoid protein being used inefficiently as an energy source. Excess protein in the diet can be metabolized for energy rather than growth, resulting in higher nitrogenous waste outputs, which can negatively affect water quality.

2.2. Lipids and Fatty Acids

Lipids serve as a concentrated source of energy and essential fatty acids (EFAs) that play vital roles in fish growth and physiological development. For fish larvae and juveniles, lipids contribute to membrane structure, hormone synthesis, and energy storage, with certain omega-3 polyunsaturated fatty acids (PUFAs) being particularly important for neural development and immune system function. Key EFAs, such as eicosapentaenoic acid (EPA), docosahexaenoic acid (DHA), and arachidonic acid (ARA), are critical for marine fish species, as they are unable to synthesize these fatty acids in sufficient quantities. DHA, for example, is crucial for the development of the brain and retina, while EPA and ARA are involved in inflammatory responses and cellular signaling pathways. Marine fish, such as seabass and seabream, have high requirements for DHA and EPA. In contrast, freshwater species may be able to synthesize some of these fatty acids from precursor molecules, but supplementation is still beneficial, especially during the early stages of development.

To ensure that larvae receive adequate amounts of these essential fatty acids, live feeds such as rotifers and Artemia are often enriched with fish oils or microalgae before being fed to the larvae. This enrichment improves the lipid composition of the live feeds, thereby ensuring that larvae obtain the DHA and EPA needed for optimal development. For juvenile fish, the inclusion of fish oil in formulated diets provides a direct source of these critical fatty acids. However, as the sustainability of fish oil comes under scrutiny due to overfishing, alternative lipid sources such as algal oil and other plant-based oils are being explored. These alternatives must be carefully balanced to ensure they meet the EFA requirements of the species being cultured.

2.3. Carbohydrates

Carbohydrates serve as an important energy source in fish diets, but their utilization by fish is generally less efficient than that of lipids or proteins. Larval and juvenile fish have a limited ability to digest and metabolize carbohydrates, primarily because of their low activity of carbohydrate-digesting enzymes such as amylase during early life stages. However, certain species, particularly omnivorous or herbivorous fish like tilapia and carp, can utilize carbohydrates more effectively as they mature.

Carbohydrates, typically in the form of starches or other digestible polysaccharides, can be included in juvenile diets to provide an additional source of energy, thus sparing protein from being used as an energy substrate. This is especially important in formulated feeds designed for species with higher tolerance to carbohydrates. Marine fish species, on the other hand, have a lower ability to utilize carbohydrates efficiently, and excessive levels of carbohydrates in their diets can impair growth performance and protein utilization. As a result, carbohydrate levels in marine fish larval diets are usually kept to a minimum to avoid metabolic imbalances. The inclusion of carbohydrates in aquafeeds must therefore be carefully managed, with species-specific considerations. Carbohydrates can also play a role in forming the physical structure of feed pellets, enhancing feed stability and water retention, particularly in juvenile diets.

2.4. Vitamins and Minerals

Vitamins and minerals are essential micronutrients that support a wide array of physiological processes in fish larvae and juveniles, including immune function, antioxidant defense, skeletal development, and metabolic regulation. These nutrients are required in minute quantities, but their deficiency can have significant impacts on growth, health, and survival.

Vitamins, particularly the fat-soluble vitamins A, D, E, and K, are crucial for a range of biological functions:

- **Vitamin A** is important for vision, immune function, and cellular differentiation.
- **Vitamin D** regulates calcium and phosphorus homeostasis, supporting bone mineralization and skeletal development.
- **Vitamin E** acts as an antioxidant, protecting cells from oxidative damage.
- **Vitamin K** is involved in blood clotting and bone metabolism.

Water-soluble vitamins such as vitamin C (ascorbic acid) and B vitamins (e.g., thiamine, riboflavin, niacin) are also essential. Vitamin C plays a pivotal role in collagen synthesis, immune response, and antioxidant protection. Its deficiency can lead to poor wound healing, skeletal deformities, and reduced disease resistance in fish larvae and juveniles.

Minerals, including calcium, phosphorus, magnesium, iron, and zinc, are required for bone formation, metabolic enzyme function, and osmotic regulation. For instance, calcium and phosphorus are integral to the formation of strong bones and scales, and an imbalance between these two minerals can lead to skeletal deformities. Zinc, meanwhile, is important for enzyme function and immune system support.

The challenge in larval and juvenile nutrition is ensuring that these vitamins and minerals are supplied in bioavailable forms that can be efficiently absorbed and utilized. In formulated diets, vitamin and mineral premixes are commonly included to provide a complete nutrient profile, while live feeds may require enrichment to address potential deficiencies.

2.5. Role of Nutritional Supplements

In addition to the basic nutrients, the use of nutritional supplements, such as probiotics, prebiotics, immunostimulants, and digestive enzymes, is becoming increasingly popular in larval and juvenile fish nutrition. These supplements can enhance nutrient utilization, promote gut health, and strengthen the immune system, leading to improved survival rates and growth performance. For example, probiotics like Bacillus and Lactobacillus species can improve gut microbiota balance, enhance nutrient absorption, and protect against pathogenic bacteria. Prebiotics, such as fructooligosaccharides (FOS), act as food for beneficial gut bacteria, promoting a healthy digestive system.

Meeting the specific nutritional requirements of larval and juvenile fish is essential for supporting their rapid growth, physiological development, and survival. By providing adequate levels of proteins, lipids, carbohydrates, vitamins, and minerals, and by utilizing nutritional supplements, aquaculture operations can improve the health and performance of fish during these early life stages. Properly formulated diets and live feed enrichment strategies are key to optimizing nutrient intake and minimizing nutritional deficiencies, leading to more sustainable and efficient aquaculture practices.

3. Feeding Strategies for Larval Fish

Feeding strategies for larval fish are crucial for ensuring optimal growth, survival, and development during the early life stages. At this stage, larval fish are highly vulnerable to environmental stressors, nutritional deficiencies, and predation, making appropriate feeding approaches essential for reducing mortality and improving health. Due to their underdeveloped digestive systems, larval fish have unique dietary needs that must be met with highly digestible, nutrient-rich feeds. The following sections detail the various feeding strategies and approaches utilized in aquaculture to meet the nutritional requirements of larval fish.

3.1. Live Feeds

Live feeds are the foundation of larval nutrition in aquaculture. These feeds are highly palatable, easily digested, and packed with essential nutrients that cater to the specific metabolic needs of larval fish. Common live feeds include microalgae, rotifers, copepods, and *Artemia nauplii*.

- Microalgae are typically used as a primary feed for zooplankton (which are then consumed by fish larvae) or directly fed to filter-feeding larvae. Microalgae are rich in essential fatty acids (EFAs), pigments, vitamins, and minerals, supporting the health and growth of both zooplankton and fish larvae.
- Rotifers (Brachionus spp.) are widely used in the early stages of larval fish culture due to their small size, slow swimming behavior, and ability to carry essential nutrients. Rotifers can be easily cultured and enriched with nutritional supplements, including EFAs, to improve their nutritional value.
- *Artemia nauplii* are one of the most widely used live feeds for fish larvae. They provide a rich source of proteins, lipids, and essential fatty acids like docosahexaenoic acid (DHA) and eicosapentaenoic acid (EPA), crucial for larval development. Artemia can also be enriched with nutrients, particularly omega-3 fatty acids, to enhance the nutritional quality for fish larvae.

One of the key advantages of live feeds is their ability to be nutrient-enriched before being offered to fish larvae. Enrichment can involve supplementation with vitamins, minerals, fatty acids, and pigments, ensuring that live feeds meet the specific nutritional needs of the target species. Enrichment with EFAs like DHA and EPA is especially important for marine fish larvae, as these species have limited ability to synthesize these essential fatty acids on their own. The palatability of live feeds and their movement in the water stimulates the larval fish's predatory instincts, encouraging feeding and improving intake rates. The high digestibility of live feeds also means that larval fish can efficiently extract nutrients, supporting rapid growth and immune function during this critical period.

3.2. Copepods and Zooplankton

Copepods and other zooplankton species are gaining increasing attention as live feeds for marine fish larvae due to their superior nutritional profiles. Copepods are naturally rich in essential nutrients that are critical for larval fish development, including phospholipids, DHA, and pigments like astaxanthin.

- Phospholipids, found abundantly in copepods, are crucial for cell membrane integrity and proper metabolic function in larvae. Their availability in the diet enhances energy metabolism and promotes growth.
- DHA, one of the most important omega-3 fatty acids, supports neural development and cognitive function in fish larvae. Copepods are an excellent source of DHA, and their inclusion in larval diets helps improve survival rates and feed efficiency.
- **Astaxanthin**, a red carotenoid pigment found in copepods, functions as an antioxidant, protecting larval fish from oxidative stress and supporting the immune system. This pigment is also involved in pigmentation, which may be crucial for species where coloration plays a role in survival.

Copepods also exhibit favorable size and behavior characteristics, making them particularly attractive to larval fish. Their small size and erratic swimming movements stimulate larval feeding responses, leading to higher ingestion rates and improved growth. This makes copepods a superior alternative to traditional live feeds like rotifers or Artemia, particularly for marine species with high nutritional requirements. Moreover, copepods' natural diet of microalgae and their rich fatty acid profile make them a more sustainable and nutritionally complete feed option. Research has shown that the inclusion of copepods in larval diets results in better growth performance, higher survival rates, and improved larval quality compared to other live feed options.

3.3. Microalgae

Microalgae play a dual role in larval nutrition: they are used as a direct feed source for filter-feeding larvae and as a feed for zooplankton species like rotifers and copepods. The nutritional content of microalgae varies between species, but they are generally rich in proteins, essential fatty acids, vitamins, and pigments, making them a vital component of the larval feed chain.

- Isochrysis, Tetraselmis, and Chlorella are among the most commonly used microalgae species in aquaculture. Isochrysis, for example, is rich in DHA, making it an excellent feed for enriching rotifers or copepods, which are then used as feed for fish larvae.

- Microalgae also provide high levels of carotenoids, such as beta-carotene and lutein, which function as antioxidants and immune modulators in larval fish. These pigments contribute to the overall health and pigmentation of fish, protecting them from oxidative stress and supporting their immune systems.

When microalgae are used directly as feed for filter-feeding fish larvae, they provide not only a rich source of nutrients but also aid in the proper functioning of the larvae's digestive system by stimulating the production of digestive enzymes. This promotes better nutrient absorption and contributes to overall larval growth and survival. Additionally, microalgae can improve water quality in the larval rearing environment by helping to stabilize water parameters and reduce harmful ammonia concentrations through their role in the nitrogen cycle.

3.4. Formulated Larval Diets

The development of formulated diets for larval fish has been a significant advancement in reducing the reliance on live feeds. Formulated diets are designed to provide the necessary balance of nutrients required by fish larvae while offering more consistency and ease of use compared to live feeds. These diets typically include a combination of proteins, lipids, carbohydrates, vitamins, and minerals tailored to the species being cultured.

- **Microencapsulation** technology is often used in the production of formulated larval diets, protecting sensitive nutrients like vitamins and essential fatty acids from oxidation and degradation during storage. Microencapsulated diets also ensure that nutrients are delivered efficiently to the larvae, promoting better growth and survival.
- **Extruded diets** are another type of formulated feed designed for larval fish. These diets are produced using high-pressure cooking processes, which improve the digestibility of feed components and reduce the presence of anti-nutritional factors.

To mimic the nutrient profile of live feeds, formulated diets often include fishmeal, fish oil, and plant proteins, along with other key ingredients. Formulated diets may also be supplemented with digestive enzymes (such as proteases and lipases) and probiotics to improve nutrient absorption and gut health. One of the main challenges with formulated larval diets is that fish larvae are highly selective feeders and may reject artificial diets if they are not appropriately sized or textured. To address this issue, formulated diets are often made in small particle sizes and designed to suspend in the water column, mimicking the movement of live prey. Recent innovations in formulated diets include the use of bioactive compounds such as immunostimulants, which enhance the larvae's immune responses, and prebiotics, which promote the growth of beneficial gut microbiota. These additives can significantly improve the health and disease resistance of larvae, reducing mortality rates and improving overall performance in aquaculture systems.

Feeding strategies for larval fish are a critical aspect of aquaculture management, directly influencing survival, growth, and overall health. Live feeds such as rotifers, copepods, and Artemia remain essential in the early stages of larval development due to their high nutritional value and ease of digestion. However, advances in formulated larval diets offer promising alternatives, providing more consistency and the potential to reduce the reliance on live feeds. By integrating both live and formulated feeds, and enhancing these with nutrient enrichment techniques, aquaculture operations can optimize larval nutrition, leading to improved survival rates and better growth performance.

4. Transition from Larval to Juvenile Diets

The transition from larval to juvenile diets represents a critical phase in the development of fish. This period is marked by significant physiological and behavioral changes, including the development of a more complex digestive system, increased growth rates, and a shift in feeding habits. Proper management of the transition period is essential to ensure that fish continue to grow optimally without suffering from stress, malnutrition, or mortality. The shift from live feeds, which are easy to digest, to formulated diets, requires careful consideration of weaning strategies and feed composition to maintain growth and health.

4.1. Weaning Strategies

Weaning is the process of gradually transitioning fish from live feeds, such as rotifers and Artemia, to **formulated diets**. This shift is challenging because fish larvae are highly dependent on live feeds during early stages, both for nutrition and to stimulate feeding behavior. A gradual weaning process ensures that larvae adapt to the new diets without adverse effects on growth, digestion, or survival.

- **Timing of Weaning:** The timing of the weaning process is critical and is species-specific. Weaning is generally initiated when the larvae have developed a sufficiently mature digestive system to handle complex formulated feeds. For many species, this occurs towards the end of the larval stage, when digestive enzymes, such as proteases and lipases, become fully functional. If weaning is started too early, fish larvae may not be able to digest the new diet effectively, resulting in malnutrition and growth suppression.

- **Introduction of Microparticulate Diets:** The use of microparticulate diets during the weaning phase is a common strategy. These diets are designed to mimic the nutritional profile and physical characteristics of live feeds. Microparticulate diets are small, buoyant, and have a texture that closely resembles live prey, making them more acceptable to larvae. By offering microparticulate feeds alongside live prey during the initial stages of weaning, fish can gradually adjust to the new diet.
- **Stepwise Reduction in Live Feeds:** Gradual reduction in live feeds is an effective strategy for minimizing stress during weaning. For example, live feed availability can be decreased over a period of several days or weeks, while formulated diets are progressively introduced. This method ensures that larvae become accustomed to the new diet and are not deprived of essential nutrients during the transition.
- **Feeding Frequency and Technique:** During the weaning period, it is important to maintain a high feeding frequency to encourage feeding behavior and provide continuous access to food. Frequent feeding also ensures that larvae have ample opportunities to consume formulated diets. Broadcast feeding or using feeding rings can help distribute feed evenly throughout the tank, ensuring that all fish have access to food.

4.2. Improving Digestibility of Formulated Feeds

The digestibility of formulated feeds is crucial during both the weaning and juvenile stages. Poorly digested feeds can lead to nutrient wastage, reduced growth, and increased susceptibility to disease. Various strategies can be employed to improve the digestibility of formulated feeds, ensuring that fish make the most efficient use of nutrients.

- **Exogenous Enzymes:** The inclusion of exogenous enzymes in formulated diets is a widely adopted strategy to enhance nutrient breakdown. For example, proteases are added to improve protein digestion, while phytases help break down phytate-bound phosphorus in plant-based ingredients, making it more bioavailable. These enzymes are particularly important in diets that contain a high proportion of plant-based proteins, as these ingredients often contain anti-nutritional factors that can hinder nutrient absorption.
- **Prebiotics and Probiotics:** The use of prebiotics and probiotics in juvenile fish diets has gained popularity due to their positive effects on gut health and digestion. Prebiotics, such as inulin and fructooligosaccharides (FOS), are non-digestible fibers that promote the growth of beneficial gut bacteria. Probiotics, including species from the genera Lactobacillus and

Bacillus, help maintain a healthy gut microbiota, improve digestion, and enhance immune function. The combination of prebiotics and probiotics in the diet, often referred to as synbiotics, can further improve nutrient assimilation and promote overall fish health.

- **Feed Processing Techniques:** Advances in feed processing, such as extrusion and microencapsulation, also play a role in improving digestibility. Extrusion techniques help to break down complex carbohydrates and increase the bioavailability of nutrients, while microencapsulation protects sensitive ingredients like vitamins and fatty acids from degradation.

Nutritional Supplements and Additives for Juvenile Fish

As fish progress to the juvenile stage, their nutritional requirements change, and it becomes increasingly important to incorporate specific supplements and additives into their diets to support growth, immune function, and overall health. Nutritional supplements and additives are used to enhance the nutritional quality of feeds, improve digestibility, and boost the fish's natural defenses against disease and environmental stressors.

5.1. Probiotics and Prebiotics

Probiotics and **prebiotics** have become essential components of modern aquaculture diets due to their beneficial effects on gut health, immune response, and disease resistance.

- **Probiotics** are live beneficial bacteria that help maintain a healthy balance of gut microbiota, which is critical for efficient digestion and nutrient absorption. Commonly used probiotic species include Lactobacillus, Bifidobacterium, and Bacillus. These probiotics outcompete harmful bacteria for nutrients and space, reducing the likelihood of gastrointestinal infections and promoting overall gut health. In addition to improving digestion, probiotics can enhance the immune system, making fish more resilient to diseases.
- **Prebiotics** are non-digestible dietary fibers that serve as a food source for beneficial gut bacteria. Prebiotics such as inulin and FOS stimulate the growth and activity of these bacteria, promoting a healthy gut microbiome. When used in combination with probiotics, prebiotics help create an environment conducive to gut health, leading to better nutrient absorption and growth performance in juvenile fish.

5.2. Digestive Enzymes

The addition of digestive enzymes to juvenile fish diets can significantly improve feed efficiency and growth rates. Enzymes such as proteases, amylases, and cellulases assist in breaking down proteins, carbohydrates, and fibers, respectively. This is particularly important in diets that contain plant-based ingredients, which can be difficult for fish to digest due to the presence of anti-nutritional factors like phytates and non-starch polysaccharides (NSPs).

- **Proteases** improve the digestion of proteins, leading to better amino acid availability and utilization for growth.
- **Amylases** enhance the breakdown of starches and other carbohydrates, providing a readily available source of energy for the fish.
- **Cellulases** break down cellulose and other complex fibers, improving the digestibility of plant-based feed ingredients and reducing waste.

By improving the digestibility of these nutrients, digestive enzymes help maximize nutrient utilization, reduce feed costs, and minimize environmental impacts through reduced nutrient excretion.

5.3. Antioxidants and Immune Enhancers

Juvenile fish are often subjected to environmental stressors such as fluctuating water quality, handling, and disease challenges. These stressors can lead to the production of reactive oxygen species (ROS), which cause oxidative stress and cellular damage. To combat this, the inclusion of antioxidants and immune enhancers in the diet is crucial for protecting fish health.

- **Antioxidants** such as vitamin E, vitamin C, and selenium play a key role in neutralizing ROS and protecting cells from oxidative damage. Vitamin E is a lipid-soluble antioxidant that protects cell membranes from lipid peroxidation, while vitamin C is a water-soluble antioxidant that regenerates vitamin E and supports immune function. Selenium, a trace mineral, is an integral part of glutathione peroxidase, an enzyme that protects cells from oxidative damage.
- **Immune Enhancers** such as beta-glucans, nucleotides, and plant-based immunostimulants can help strengthen the fish's immune system. Beta-glucans, derived from yeast cell walls, stimulate the innate immune system by activating immune cells like macrophages and neutrophils. Nucleotides play a role in cell division and immune function, helping fish recover from stress and disease. Plant-based immunostimulants, such as those derived from garlic and ginger, contain bioactive compounds that boost the fish's natural defenses against pathogens.

These additives not only improve the overall health and resilience of juvenile fish but also contribute to better growth performance and survival rates in aquaculture systems.

6. Innovative Nutritional Strategies

Innovative nutritional strategies in aquaculture focus on improving sustainability, nutritional quality, and fish health, especially during the critical larval and juvenile stages. These strategies include exploring alternative protein sources, optimizing feed formulations, and incorporating bioactive compounds that enhance immunity and stress resistance. Below is an in-depth exploration of some of the most promising approaches.

6.1. Insect-Based Proteins

The use of insect-based proteins in fish diets is gaining significant attention as a sustainable alternative to conventional protein sources like fishmeal and soybean meal. Insects, particularly species such as black soldier fly larvae (BSFL) and mealworms, offer several nutritional and environmental advantages:

- **Rich Nutritional Profile:** Insect proteins are high in essential amino acids like lysine, methionine, and leucine, which are vital for growth and muscle development in juvenile fish. Insects also provide a good source of lipids, including polyunsaturated fatty acids, which support energy needs and cell membrane function.
- **Micronutrients:** Insects contain essential micronutrients such as zinc, iron, and calcium, which contribute to overall health, immune function, and skeletal development. These micronutrients are also bioavailable, meaning they are easily absorbed by the fish.
- **Environmental Benefits:** Insect farming is highly sustainable due to the ability of insects to grow on organic waste streams, reducing environmental impact and feed production costs. They require less land, water, and feed resources compared to traditional protein sources like livestock.
- **Palatability and Growth Performance:** Studies have shown that insect-based diets can improve growth performance, feed conversion ratios (FCR), and survival rates in juvenile fish. The natural palatability of insects makes them readily accepted by fish, ensuring smooth weaning from live feeds to formulated diets.

Research is ongoing to optimize the use of insect-based proteins in aquafeeds, focusing on issues such as processing techniques, inclusion levels, and nutrient digestibility to further enhance their benefits for aquaculture.

6.2. Algal Biomass

Algal biomass is another promising feed ingredient being explored for its nutritional richness and sustainability. Microalgae such as Spirulina, Chlorella, and Nannochloropsis offer a wide array of essential nutrients that support the growth and health of larval and juvenile fish.

- **Essential Fatty Acids:** Algae are a rich source of omega-3 fatty acids, particularly eicosapentaenoic acid (EPA) and docosahexaenoic acid (DHA), which are crucial for brain development, immune response, and overall growth. These fatty acids are especially important for marine fish species that have limited ability to synthesize omega-3 fatty acids.
- **Proteins and Amino Acids:** Algal biomass provides a high-quality source of protein, often containing all essential amino acids needed for optimal fish growth. Algae like Spirulina have protein contents of up to 60-70%, making them a highly concentrated source of nutrients.
- **Pigments and Antioxidants:** Algal pigments, such as carotenoids, enhance the coloration of fish, which is important for market value, especially in ornamental species. Additionally, carotenoids like astaxanthin act as antioxidants, reducing oxidative stress and enhancing the fish's immune response.
- **Sustainability:** Algal cultivation is considered environmentally friendly, as algae grow rapidly, require fewer resources, and can be produced in both marine and freshwater environments. Additionally, algae can be cultivated using photobioreactors or open ponds, which make them a scalable option for aquaculture feed production.

The inclusion of algal biomass in aquafeeds is still being optimized, but its potential to replace a portion of fishmeal and fish oil in diets is being actively researched. Algae not only enhance nutrition but also promote sustainability in aquaculture.

6.3. Functional Feeds

Functional feeds are diets that go beyond basic nutrition by incorporating bioactive compounds that provide additional health benefits, such as improved immune function, disease resistance, and stress resilience. These diets are designed to address specific challenges in aquaculture, including disease outbreaks and environmental stressors, by enhancing the natural defenses of fish.

- **Herbal Extracts and Essential Oils:** The inclusion of plant-derived compounds such as garlic extract, oregano oil, and curcumin in aquafeeds has gained attention due to their antimicrobial, anti-inflammatory, and antioxidant properties. These compounds boost the fish's immune system, helping them resist infections and recover from stress more effectively. For example, garlic extract has been shown to enhance immune responses in various fish species, while oregano oil exhibits strong antimicrobial activity against common fish pathogens.

- **Bioactive Peptides:** Peptides derived from fish, algae, and plants are incorporated into functional feeds for their antioxidant and antimicrobial properties. These peptides help reduce the effects of oxidative stress, which can occur during periods of rapid growth or environmental challenges, thereby supporting better health and growth performance.
- **Beta-Glucans and Nucleotides:** Beta-glucans, derived from yeast cell walls, are powerful immunostimulants that enhance the fish's innate immune response by activating immune cells such as macrophages and neutrophils. Nucleotides, essential for cell division and immune function, help juvenile fish recover from stress and enhance their resistance to diseases, particularly during the weaning and juvenile stages when they are more susceptible to infections.
- **Prebiotics and Probiotics:** The inclusion of prebiotics and probiotics in functional feeds improves gut health by maintaining a balanced gut microbiota, enhancing digestion, and promoting nutrient absorption. This approach not only improves feed efficiency but also boosts immunity and growth.

Functional feeds represent a cutting-edge approach to aquaculture nutrition, combining the benefits of conventional nutrients with bioactive compounds that promote health and performance.

Table: Nutritional Strategies to Improve Larval and Juvenile Fish Survival

Nutritional Component	Description	Role in Larval and Juvenile Fish	Examples/Sources	Benefits
Protein and Amino Acids	Fundamental macronutrient for growth and development	Supports muscle development, growth, and immune function	Fishmeal, soybean meal, rotifers, Artemia, insect protein (black soldier fly larvae, mealworms)	High growth rates, optimal immune responses
Essential Fatty Acids (EFAs).	Lipids critical for energy and neural development	Important for brain development, membrane function, and immune response	Fish oil, marine oils, algae, copepods	Improves survival, immune function, and neural development
Carbohydrates	Supplemental energy source	Reduces protein used for energy and supports metabolism	Starches, plant-based ingredients	Energy source for juvenile fish, enhances growth without over-reliance on protein
Vitamins and Minerals	Micronutrients essential for metabolic processes	Crucial for antioxidant defense, bone formation, and immune function	Vitamin C, Vitamin E, Vitamin A, calcium, phosphorus, zinc	Supports skeletal development, enhances immune response, prevents deformities
Live Feeds	Natural feeds rich in nutrients	Highly digestible and palatable for early-stage larvae	Microalgae, rotifers, Artemia, copepods	Provides complete nutrition, enhances survival and growth
Copepods and Zooplankton	High-quality live feed with superior nutritional value	Rich in DHA, EPA, and pigments important for larval development	Marine copepods, zooplankton species	Improves survival, enhances pigmentation, supports early development
Microalgae	Feed source and water quality enhancer	Provides proteins, fatty acids, and pigments (carotenoids)	Isochrysis, Tetraselmis, Chlorella, Spirulina	Supports immune function, growth, and pigmentation
Formulated Diets	Artificial feeds developed for specific life stages	Designed to meet the nutritional requirements of larvae and juveniles	Microencapsulated diets, extruded feeds, algae-based feeds	Mimics live feed nutrition, reduces reliance on live feeds, promotes gut health

Probiotics and Prebiotics	Additives to improve gut health and digestion	Enhances immune system, improves nutrient assimilation	Lactobacillus, Bacillus species, non-digestible carbohydrates (FOS, MOS)	Improves survival, reduces disease incidence, promotes growth
Digestive Enzymes	Enzyme supplementation to improve nutrient digestion	Increases nutrient bioavailability, especially in plant-based diets	Protease, amylase, cellulase, phytase	Enhances feed efficiency, improves growth and feed conversion ratio (FCR)
Antioxidants and Immune Enhancers	Nutritional compounds for stress resistance and immune function	Protects fish from oxidative stress, boosts immune responses	Vitamin E, Vitamin C, selenium, beta-glucans, nucleotides	Reduces mortality, enhances disease resistance, supports overall health
Insect-Based Protein	Sustainable protein alternative	Provides essential amino acids, lipids, and micronutrients	Black soldier fly larvae, mealworms	Enhances growth, sustainable feed ingredient, reduces environmental impact
Algal Biomass	Sustainable alternative protein and nutrient source	Provides essential fatty acids, vitamins, and pigments	Spirulina, Chlorella, Dunaliella	Supports immune function, growth, and pigmentation
Functional Feeds	Feeds with bioactive compounds for health benefits	Enhances growth, immunity, and stress resistance	Herbal extracts, essential oils, bioactive peptides	Improves resilience to stress, promotes health, and improves survival
Weaning Strategies	Gradual transition from live to formulated feeds	Ensures smooth transition without compromising growth	Microparticulate diets, weaning over several days	Reduces mortality, maintains growth, improves feed acceptance

7. Conclusion

Nutritional strategies play a pivotal role in the early stages of fish development, ensuring optimal growth, survival, and health. As aquaculture continues to expand, the adoption of innovative feed ingredients and advanced feeding technologies will be essential to support sustainable growth. The use of insect-based proteins and algal biomass offers promising alternatives to traditional feed ingredients, reducing the reliance on fishmeal and soybean meal while maintaining high nutritional quality. Meanwhile, functional feeds that incorporate bioactive compounds provide additional health benefits, enhancing immunity and stress resilience in juvenile fish.

Further research into species-specific nutritional requirements and the development of cost-effective, sustainable feed solutions will continue to shape the future of aquaculture. By optimizing early-stage nutrition, the industry can improve survival rates, growth performance, and overall fish health, contributing to the long-term success and sustainability of global aquaculture operations.

References

Balcázar, J. L., de Blas, I., Ruiz-Zarzuela, I., Cunningham, D., Vendrell, D., & Múzquiz, J. L. (2006). The role of probiotics in aquaculture. Veterinary Microbiology, 114(3-4), 173-186.

Boglione, C., Gisbert, E., Gavaia, P., Witten, P. E., Moren, M., Fontagné, S., & Koumoundouros, G. (2013). Skeletal anomalies in reared European fish larvae and juveniles. Part 2: Main typologies, occurrences, and causative factors. Reviews in Aquaculture, 5(S1), S121-S167.

Cahu, C. L., Zambonino Infante, J. L., & Barbosa, V. (2003). Effect of diet on pre-larval development in marine fish. Aquaculture, 227(1-4), 245-258.

Conceição, L. E. C., Morais, S., & Rønnestad, I. (2007). Tracing larval fish digestion: Development and application of bioanalytical tools in larval fish studies. Aquaculture, 268(1-4), 82-91.

Conceição, L. E. C., Yufera, M., Makridis, P., Morais, S., & Dinis, M. T. (2010). Live feeds for early stages of fish rearing. Aquaculture Research, 41(5), 613-640.

Faulk, C. K., & Holt, G. J. (2008). Advances in rearing red drum (Sciaenops ocellatus) larvae: A review. Aquaculture, 277(3-4), 168-180.

Gisbert, E., & Moyano, F. J. (2007). Synergistic effects of diet and digestive function in fish larvae. Aquaculture, 268(1-4), 24-34.

Hamre, K., Yúfera, M., Rønnestad, I., Boglione, C., Conceição, L. E., & Izquierdo, M. (2013). Fish larval nutrition and feed formulation: Knowledge gaps and bottlenecks for advances in larval rearing. Reviews in Aquaculture, 5(S1), S26-S58.

Hart, S. D., & Morais, S. (2020). Influence of diet composition and feeding regime on growth and digestive enzyme activities in marine fish larvae. Aquaculture Nutrition, 26(5), 1853-1867.

Holt, G. J. (2011). Larval fish nutrition. John Wiley & Sons.

Izquierdo, M. S., Fernandez-Palacios, H., & Tacon, A. G. (2001). Effect of broodstock nutrition on reproductive performance of fish. Aquaculture, 197(1-4), 25-42.

Kolkovski, S. (2001). Digestive enzymes in fish larvae and juveniles—Implications and applications to formulated diets. Aquaculture, 200(1-2), 181-201.

Koven, W., Kolkovski, S., Tandler, A., Kissil, G., & Applebaum, S. (1998). The effect of dietary phosphatidylcholine on the assimilation and metabolism of lipids in gilthead seabream (Sparus aurata) larvae. Fish Physiology and Biochemistry, 18(1), 69-77.

Lall, S. P., & Lewis-McCrea, L. M. (2007). Role of nutrients in skeletal metabolism and pathology in fish—An overview. Aquaculture, 267(1-4), 3-19.

Li, P., & Gatlin III, D. M. (2004). Dietary brewer's yeast and the prebiotic Grobiotic™ AE influence growth performance, immune responses and resistance of hybrid striped bass (Morone chrysops×M. saxatilis) to Streptococcus iniae infection. Aquaculture, 231(1-4), 445-456.

Makridis, P., & Olsen, Y. (1999). Protein depletion and accumulation during the development of Artemia in the presence and absence of food. Marine Biology, 134(2), 335-342.

Olivotto, I., Planas, M., Simões, N., Holt, G. J., Avella, M. A., & Calado, R. (2011). Advances in breeding and rearing marine ornamentals. Journal of the World Aquaculture Society, 42(2), 135-166.

Olsen, R. E., & Ringø, E. (1998). Lipid digestibility in fish: A review. Aquaculture Nutrition, 4(3), 189-198.

Rønnestad, I., Yúfera, M., Ueberschär, B., Ribeiro, L., Sæle, Ø., & Boglione, C. (2013). Feeding behaviour and digestive physiology in larval fish: Current knowledge, and gaps and bottlenecks in research. Reviews in Aquaculture, 5(S1), S59-S98.

Sales, J., & Janssens, G. P. (2003). Nutrient requirements of ornamental fish. Aquatic Living Resources, 16(6), 533-540.

Santos, G. A., Nematipour, G. R., & Gatlin III, D. M. (1997). Dietary thiamin requirement of juvenile striped bass Morone chrysops × M. saxatilis. Journal of Nutrition, 127(7), 1400-1405.

Shields, R. J., Bell, J. G., Luizi, F. S., Gara, B., Bromage, N. R., & Sargent, J. R. (1999). Natural copepods are superior to enriched Artemia nauplii as feed for halibut larvae (Hippoglossus hippoglossus) in terms of survival, pigmentation, and retinal morphology: Relation to dietary essential fatty acids. Journal of Nutrition, 129(6), 1186-1194.

Sorgeloos, P., Dhert, P., & Candreva, P. (2001). Use of the brine shrimp, Artemia spp., in marine fish larviculture. Aquaculture, 200(1-2), 147-159.

Tandler, A., & Kolkovski, S. (1991). Rates of digestion and stomach evacuation of live food (Artemia nauplii) in seabream (Sparus aurata) larvae under various feeding conditions. Aquaculture, 92(1-2), 185-192.

Treece, G. D. (2000). Artemia production for marine larval fish culture: A review. Aquaculture, 200(1-2), 175-188.

Vadstein, O., Øie, G., Olsen, Y., & Reinertsen, H. (2001). A strategy to enhance production yield in marine larviculture using biocontrol based on the colonization of bacteria. Aquaculture, 200(1-2), 231-245.

Van Anholt, R. D., Koven, W. M., Lutzky, S., & Rønnestad, I. (2004). Copepods as live feed in marine fish larviculture: A review. Aquaculture Nutrition, 10(6), 434-451.

Yúfera, M., Pascual, E., & Fernández-Díaz, C. (1999). A comparative study of the effectiveness of four Artemia strains for larval development of the gilthead seabream (Sparus aurata L.). Aquaculture, 174(3-4), 337-344.

Watanabe, T., Kiron, V., & Satoh, S. (1997). Trace minerals in fish nutrition. Aquaculture, 151(1-4), 185-207.

Yufera, M., & Darias, M. J. (2007). The onset of exogenous feeding in marine fish larvae. Aquaculture, 268(1-4), 53-63.

9

Antioxidants in Aquaculture: Protecting Fish from Oxidative Stress

1. Introduction

Aquaculture has emerged as a vital sector in global food production, with the increasing demand for fish and seafood as a primary source of protein for a rapidly growing population. According to the Food and Agriculture Organization (FAO), aquaculture contributes significantly to global fisheries production, accounting for nearly 50% of the total fish consumed worldwide. This remarkable growth trajectory is driven by the need to meet food security challenges while providing sustainable and nutritious food sources. Despite its importance, intensive aquaculture practices present numerous challenges that can adversely affect fish health and productivity. Fish raised in aquaculture systems are often subjected to a variety of environmental stressors, including poor water quality, high stocking densities, rapid growth rates, and fluctuations in temperature. These stressors lead to increased metabolic activity in fish, which can elevate the production of reactive oxygen species (ROS) — highly reactive molecules that can cause cellular damage. Under normal physiological conditions, the production of ROS is counterbalanced by the antioxidant defense mechanisms in fish. However, when the production of ROS exceeds the capacity of these defense systems, oxidative stress ensues.

Oxidative stress is a detrimental condition characterized by an imbalance between ROS generation and the antioxidant capacity of an organism. In fish, this condition can manifest in various forms, including cellular damage, lipid peroxidation, protein oxidation, and DNA damage. The consequences of oxidative stress are severe, resulting in impaired growth, compromised immune responses, increased susceptibility to diseases, and ultimately higher mortality rates. Understanding the underlying mechanisms of oxidative stress in aquaculture is critical for developing effective strategies to mitigate its impact and improve fish welfare.

Antioxidants play a fundamental role in protecting aquatic organisms from oxidative damage. They are compounds that can neutralize ROS, thereby preventing cellular injury and maintaining redox homeostasis. The antioxidant defense system in fish comprises both enzymatic and non-enzymatic components. Enzymatic antioxidants, such as superoxide dismutase (SOD), catalase, and glutathione peroxidase (GPx), work to convert harmful ROS into less reactive species. Non-enzymatic antioxidants, including vitamins C and E, carotenoids, and polyphenols, can scavenge free radicals and reduce oxidative stress.

This chapter delves into the multifaceted role of antioxidants in aquaculture, highlighting their importance in promoting fish health, growth performance, and resilience against environmental stressors. We will explore the various sources of antioxidants, their mechanisms of action, and the potential benefits they offer to fish in aquaculture settings. Additionally, we will discuss practical applications for incorporating antioxidant supplements into fish diets, emphasizing their role in enhancing overall health and productivity in aquaculture systems.

As the aquaculture industry continues to expand and evolve, addressing the challenges posed by oxidative stress through effective nutritional strategies will be crucial for ensuring sustainable and successful fish farming practices. By enhancing our understanding of antioxidants and their application in aquaculture, we can foster healthier fish populations, improve food security, and contribute to the overall sustainability of aquatic food production systems.

2. Understanding Oxidative Stress in Fish

2.1. Reactive Oxygen Species (ROS) and Their Impact

Reactive oxygen species (ROS) are highly reactive molecules that contain oxygen and are generated during normal cellular metabolism. These species include free radicals, such as superoxide anions (O2−) and hydroxyl radicals (OH•), as well as non-radical species like hydrogen peroxide (H2O2). ROS are primarily produced in the mitochondria during aerobic respiration, where they are formed as byproducts of the electron transport chain. While low levels of ROS are essential for various physiological functions, including cell signaling, immune response, and the regulation of metabolic pathways, excessive accumulation of these molecules can lead to oxidative stress.

In fish, oxidative stress is a condition that arises when there is an imbalance between ROS production and the organism's antioxidant defenses. This imbalance can result in significant damage to cellular components, including lipids, proteins, and nucleic acids. Lipid peroxidation, a common consequence of oxidative stress, leads to the degradation of cell membranes, altering their integrity and function. Protein oxidation can impair enzyme activity and disrupt cellular processes, while DNA damage can lead to mutations and affect cell proliferation. The detrimental effects of oxidative stress on fish health are profound. Chronic oxidative stress can lead to a range of health issues, including impaired immune function, slower growth rates, and reduced feed efficiency. Fish suffering from oxidative stress are more susceptible to infections and diseases due to compromised immune responses. Furthermore, oxidative stress can increase mortality rates, posing significant challenges to aquaculture operations. Consequently, understanding and managing oxidative stress is crucial for maintaining the health of fish populations and optimizing aquaculture production.

2.2. Sources of Oxidative Stress in Aquaculture

Fish in aquaculture systems are exposed to various environmental and management-related stressors that can induce oxidative stress. Some of the key sources of oxidative stress in aquaculture include:

1. **High Stocking Densities**: Intensive aquaculture practices often involve high stocking densities, which can lead to increased competition for resources, heightened aggression, and physical stress among fish. The resulting stress can elevate metabolic rates, causing an overproduction of ROS and overwhelming the fish's antioxidant defenses.
2. **Poor Water Quality**: The quality of water in aquaculture systems plays a critical role in fish health. Elevated levels of ammonia, nitrite, and other pollutants can cause oxidative stress by directly damaging cellular structures or by inducing inflammation. Poor water quality can impair fish growth, increase stress levels, and heighten susceptibility to diseases.
3. **Nutritional Deficiencies**: The nutritional composition of fish diets is pivotal for maintaining health and resilience. Diets lacking essential nutrients, particularly antioxidants like vitamins E and C, can diminish the fish's ability to neutralize ROS. Consequently, fish with poor nutritional status are more vulnerable to oxidative stress, leading to reduced growth rates and increased disease susceptibility.
4. **Thermal Stress**: Temperature fluctuations or suboptimal rearing temperatures can significantly affect fish metabolism. Elevated temperatures can increase metabolic rates, leading to heightened ROS production. Conversely, low temperatures can impair enzyme activity and physiological processes, potentially causing stress and oxidative damage.
5. **Diseases and Infections**: The presence of pathogens, such as bacteria and viruses, can elicit an immune response in fish. While this response is

necessary for combating infections, it can also lead to the overproduction of ROS. The inflammatory process associated with fighting infections can further exacerbate oxidative stress, impacting overall fish health and survival.

6. **Handling and Transportation**: Stress induced during handling and transportation, including crowding and physical injury, can lead to a surge in ROS production. This can compromise the health and welfare of fish, making them more susceptible to diseases and reducing their viability in aquaculture settings.

3. Antioxidants: Definition and Mechanisms of Action

Antioxidants are essential molecules that inhibit the oxidation of other compounds by neutralizing reactive oxygen species (ROS) and free radicals. They serve as a critical defense mechanism to protect aquatic organisms, particularly fish, from oxidative damage induced by environmental stressors, metabolic processes, and diseases. By maintaining the balance between ROS production and antioxidant defenses, antioxidants play a vital role in supporting the overall health and vitality of fish in aquaculture systems. Antioxidants can be broadly classified into two main categories: enzymatic and non-enzymatic antioxidants.

3.1. Enzymatic Antioxidants

Enzymatic antioxidants are naturally occurring enzymes produced within cells that perform critical functions in detoxifying ROS. These enzymes catalyze biochemical reactions that convert harmful ROS into less toxic molecules, thereby reducing oxidative damage and maintaining cellular homeostasis. The key enzymatic antioxidants include:

- **Superoxide Dismutase (SOD)**
  - **Function**: SOD is the first line of defense against oxidative stress. It catalyzes the conversion of superoxide anion (O2−), a highly reactive free radical, into hydrogen peroxide (H2O2) and molecular oxygen (O2). This conversion is crucial because superoxide can initiate a cascade of oxidative reactions that lead to significant cellular damage.
  - **Mechanism**: SOD operates through a catalytic cycle involving the transfer of electrons. It contains metal cofactors, typically copper, zinc, or manganese, which facilitate the conversion of superoxide to hydrogen peroxide. By efficiently removing superoxide from the cellular environment, SOD helps prevent damage to proteins, lipids, and DNA.

- **Catalase**
  - **Function**: Catalase further detoxifies hydrogen peroxide, converting it into water (H2O) and oxygen (O2). This is essential as hydrogen peroxide can lead to the formation of highly reactive hydroxyl radicals, which are particularly damaging to cellular structures.
  - **Mechanism**: Catalase operates through a two-step process: in the first step, it binds to hydrogen peroxide, and in the second step, it catalyzes the decomposition of hydrogen peroxide into water and oxygen. This rapid detoxification prevents the accumulation of hydrogen peroxide and protects the cell from oxidative damage.
- **Glutathione Peroxidase (GPx)**
  - **Function**: GPx is another key enzymatic antioxidant that reduces lipid hydroperoxides and hydrogen peroxide, thus protecting cell membranes and other cellular components from oxidative damage.
  - **Mechanism**: GPx utilizes glutathione (GSH), a vital tripeptide antioxidant, as a substrate. In this process, GPx catalyzes the reduction of hydrogen peroxide and lipid hydroperoxides, converting GSH to oxidized glutathione (GSSG). The regeneration of GSH is critical for sustaining the antioxidant defense system within the cell.

These enzymatic antioxidants work synergistically to provide a robust defense against oxidative stress, ensuring the maintenance of cellular health and function.

3.2. Non-Enzymatic Antioxidants

Non-enzymatic antioxidants are compounds that can be obtained through diet or synthesized by the organism. They play a crucial role in neutralizing ROS and protecting cellular structures through various mechanisms. Key non-enzymatic antioxidants include:

- **Vitamin C (Ascorbic Acid)**
  - **Function**: Vitamin C is a water-soluble antioxidant that scavenges ROS in aqueous environments, such as blood plasma and cytosol, thereby protecting tissues from oxidative damage.
  - **Mechanism**: Ascorbic acid donates electrons to neutralize free radicals, effectively reducing their reactivity. It also plays a role in regenerating other antioxidants, such as vitamin E, by converting oxidized forms back to their active states. This recycling effect amplifies the overall antioxidant capacity within the organism.

- **Vitamin E (α-Tocopherol)**
    - **Function**: Vitamin E is a lipid-soluble antioxidant that is essential for protecting cell membranes from lipid peroxidation. It integrates into the lipid bilayer of cell membranes, where it can intercept free radicals and prevent oxidative damage.
    - **Mechanism**: By donating hydrogen atoms to free radicals, vitamin E stabilizes the radicals, thereby preventing the chain reaction of lipid peroxidation. This protective action is crucial in maintaining membrane integrity and function, particularly in high-fat tissues.
- **Carotenoids**
    - **Function**: Carotenoids, such as astaxanthin and β-carotene, are pigments found in various aquatic organisms. They exhibit potent antioxidant properties and play a role in enhancing pigmentation in fish, which can improve their market value.
    - **Mechanism**: Carotenoids act by quenching singlet oxygen and scavenging free radicals. Their unique structure allows them to absorb excess energy and neutralize reactive species, thus protecting cellular components from oxidative damage. In addition to their antioxidant properties, carotenoids also contribute to immune function and overall health.
- **Polyphenols**
    - **Function**: Polyphenols are plant-derived compounds abundant in fruits, vegetables, herbs, and algae. They exhibit strong antioxidant activity and contribute to the health benefits of dietary supplements and functional feeds in aquaculture.
    - **Mechanism**: Polyphenols neutralize free radicals through various mechanisms, including electron donation, metal chelation, and modulation of signaling pathways related to oxidative stress. They may also enhance the expression of antioxidant enzymes, contributing to the organism's overall antioxidant defense capacity.
- **Glutathione (GSH)**
    - **Function**: GSH is a critical tripeptide involved in the detoxification of ROS and the maintenance of other antioxidants in their active form. It serves as a key substrate for enzymatic antioxidants like GPx and plays a vital role in cellular redox homeostasis.
    - **Mechanism**: GSH can directly scavenge free radicals and reactive species. Additionally, it participates in the reduction of oxidized forms of other antioxidants, ensuring that the cellular antioxidant

pool remains functional. The levels of GSH are closely linked to the overall health and resilience of fish against oxidative stress.

4. The Role of Antioxidants in Aquaculture

Antioxidants play a critical role in maintaining fish health and optimizing aquaculture production by counteracting oxidative stress caused by various environmental and physiological factors. Their multifunctional benefits encompass protection against lipid peroxidation, enhancement of immune function, improvement of growth and feed efficiency, and increased stress resistance. Understanding these roles is vital for developing effective nutritional strategies that incorporate antioxidants into fish diets.

4.1. Protection Against Lipid Peroxidation

Lipid peroxidation is a detrimental process involving the oxidative degradation of lipids, primarily within cell membranes. This process can initiate a cascade of cellular damage, resulting in the formation of toxic byproducts, including malondialdehyde (MDA) and 4-hydroxynonenal (4-HNE). These byproducts can compromise membrane integrity, alter membrane fluidity, and ultimately lead to cell death.

- **Mechanism of Lipid Peroxidation**
  - Lipid peroxidation typically begins with the attack of ROS on polyunsaturated fatty acids (PUFAs) present in cell membranes. The reaction produces lipid free radicals, which propagate further oxidation by reacting with other lipid molecules, creating a chain reaction.
  - This process not only damages cell membranes but can also disrupt cellular signaling pathways, impairing the overall function of the cell and the organism.
- **Role of Antioxidants**
  - **Vitamin E (α-Tocopherol)**: As a primary lipid-soluble antioxidant, vitamin E integrates into cell membranes, where it protects against lipid peroxidation by donating hydrogen atoms to free radicals. This action stabilizes the radicals, preventing them from reacting with other lipid molecules.
  - **Carotenoids**: Compounds such as astaxanthin and β-carotene are known to quench singlet oxygen and scavenge free radicals, reducing the risk of lipid peroxidation. They help maintain the structural integrity of cell membranes, supporting cellular functions and overall health.

- By preventing lipid peroxidation, antioxidants help to maintain membrane fluidity and functionality, which is critical for nutrient transport, signal transduction, and overall cellular health.

4.2. Enhancing Immune Function

Oxidative stress has a profound impact on the immune system of fish, compromising their ability to respond effectively to infections and diseases. Antioxidants play a crucial role in supporting and enhancing immune function in several ways:

- **Protection of Immune Cells**
  - Antioxidants protect immune cells, such as leukocytes and macrophages, from oxidative damage. By neutralizing excess ROS, antioxidants prevent cellular apoptosis and ensure the survival and efficacy of immune cells in combating pathogens.
- **Promotion of Immune-Related Proteins**
  - Antioxidants, particularly vitamin C, are essential for the synthesis of immune-related proteins, such as antibodies and cytokines. Vitamin C enhances the activity of immune cells, leading to a more robust immune response.
  - **Collagen Production**: Vitamin C is vital for the synthesis of collagen, an important structural protein in connective tissues. Collagen is crucial for wound healing and tissue repair, allowing fish to recover from injuries and infections more effectively.
- **Reduction of Inflammation**
  - Antioxidants help to modulate inflammatory responses, which can be exacerbated by oxidative stress. By reducing inflammation, antioxidants support the overall health and well-being of fish, minimizing the impact of diseases and infections.

4.3. Improving Growth and Feed Efficiency

Oxidative stress can negatively affect growth rates and feed efficiency in fish by impairing essential metabolic processes, such as protein synthesis and energy metabolism.

- **Impact on Protein Synthesis**
  - High levels of ROS can damage proteins and nucleic acids, leading to decreased protein synthesis and disrupted cellular functions. This can result in slower growth rates and reduced feed conversion ratios (FCR).

- **Antioxidant Mitigation**
  - The inclusion of antioxidants in fish diets has been shown to counteract oxidative damage, thereby promoting normal protein synthesis and enhancing growth performance. For instance, dietary vitamin E has been linked to improved growth rates in fish, particularly in species raised under stressful conditions.
- **Feed Efficiency**
  - Antioxidants contribute to improved feed efficiency by enhancing nutrient utilization and energy metabolism. By reducing oxidative damage, antioxidants help ensure that nutrients from feeds are effectively absorbed and utilized for growth, thereby enhancing FCR.

4.4. Stress Resistance

Fish are frequently exposed to various environmental stressors, including poor water quality, temperature fluctuations, and handling stress. These factors can lead to increased ROS production and heightened oxidative stress, compromising fish health and survival.

- **Environmental Stressors**
  - **Water Quality**: Elevated levels of ammonia, nitrite, and other pollutants can induce oxidative stress in fish, leading to compromised health and increased mortality.
  - **Temperature Fluctuations**: Sudden changes in temperature can elevate metabolic rates and ROS production, contributing to oxidative stress.
  - **Handling Stress**: Fish handling during harvesting or stocking can induce stress, further increasing ROS levels.
- **Role of Antioxidants in Stress Resistance**
  - Antioxidant supplementation in fish diets can enhance stress resilience by reducing ROS levels and preventing cellular damage. This support helps fish cope with environmental challenges, leading to improved survival rates and overall health.
  - Studies have shown that dietary antioxidants can enhance the stress tolerance of fish species, allowing them to maintain physiological functions and reduce mortality rates under adverse conditions.

5. Antioxidant Supplements in Fish Diets

Antioxidants play a vital role in supporting the health and productivity of fish in aquaculture. Due to their inability to synthesize certain antioxidants endogenously, many fish species rely on dietary supplementation to meet their antioxidant needs. This section delves into various antioxidant supplements commonly used in fish diets, their mechanisms of action, and their benefits for fish health and growth.

5.1. Vitamin C

Importance

Vitamin C (ascorbic acid) is an essential water-soluble antioxidant that plays several critical roles in fish physiology. Notably, most fish species cannot synthesize vitamin C, making it essential to include it in their diets. Its primary functions include:

- **Collagen Synthesis**: Vitamin C is crucial for the hydroxylation of proline and lysine residues in collagen, contributing to the structural integrity of connective tissues.
- **Immune Function**: It enhances the activity of immune cells, promoting the synthesis of antibodies and cytokines, which are vital for the immune response.
- **Protection Against Oxidative Stress**: As a potent scavenger of reactive oxygen species (ROS), vitamin C mitigates oxidative stress, helping to maintain cellular health.

Dietary Supplementation

Vitamin C is often included in fish diets as ascorbate-2-monophosphate (AMP), which improves its stability during feed processing. Studies have demonstrated that adequate vitamin C supplementation leads to:

- **Improved Growth**: Fish receiving sufficient vitamin C show enhanced growth rates, likely due to improved collagen formation and overall health.
- **Increased Survival**: Vitamin C supplementation can significantly enhance survival rates in various species, including **rainbow trout** (*Oncorhynchus mykiss*), **tilapia** (*Oreochromis spp.*), and **common carp** (*Cyprinus carpio*).
- **Enhanced Disease Resistance**: Fish supplemented with vitamin C have shown increased resistance to infectious diseases, likely due to strengthened immune responses.

5.2. Vitamin E

Importance

Vitamin E (α-tocopherol) is a fat-soluble antioxidant that plays a critical role in protecting cell membranes from oxidative damage caused by ROS. It works synergistically with vitamin C by regenerating oxidized vitamin E, thereby enhancing its antioxidant capacity.

Dietary Supplementation

Vitamin E is particularly important in fish diets, especially for species with high-fat content, such as salmonids and marine fish. The benefits of vitamin E supplementation include:

- **Reduction in Lipid Peroxidation**: Vitamin E effectively prevents the oxidative degradation of lipids in cell membranes, reducing the formation of harmful byproducts associated with lipid peroxidation.
- **Improved Immune Function**: Adequate vitamin E levels bolster immune responses in fish, enhancing resistance to diseases and infections.
- **Enhanced Growth Performance**: Studies have shown that vitamin E supplementation improves growth rates in various fish species by promoting efficient nutrient utilization and maintaining cellular integrity.

5.3. Carotenoids

Importance

Carotenoids are natural pigments found in plants and algae that possess potent antioxidant properties. Two significant carotenoids used in aquaculture are astaxanthin and **β-carotene**.

Dietary Supplementation

Carotenoids offer several benefits in fish diets:

- **Stress Resistance**: Astaxanthin, in particular, has been shown to enhance the stress tolerance of fish by reducing oxidative damage and supporting immune function.
- **Immune Function**: Carotenoids help modulate immune responses, making fish more resilient to infections and diseases.
- **Growth Enhancement**: Supplementing fish diets with carotenoids has been associated with improved growth performance in various species.
- **Pigmentation**: Astaxanthin is responsible for the red-orange coloration in species like salmon and shrimp, adding economic value to aquaculture products.

5.4. Polyphenols and Herbal Extracts

Importance

Polyphenols are plant-derived compounds known for their strong antioxidant properties. They are found in various natural sources, including plants, algae, and herbal extracts.

Dietary Supplementation

Polyphenols offer multiple benefits in aquaculture:

- **Antioxidant Activity**: Polyphenols exhibit significant free radical scavenging activity, helping to mitigate oxidative stress in fish.
- **Anti-inflammatory Effects**: Many polyphenols have anti-inflammatory properties that can benefit fish health by reducing tissue inflammation associated with stress and infections.
- **Growth Promotion**: Extracts from plants such as **green tea, turmeric**, and **rosemary** have been shown to enhance growth and improve feed efficiency in fish.
- **Immune Response Improvement**: Supplementing fish diets with polyphenol-rich extracts can enhance immune responses and promote overall health.

5.5. Synthetic Antioxidants

Importance

Synthetic antioxidants, such as butylated hydroxytoluene (BHT) and butylated hydroxyanisole (BHA), are often incorporated into aquaculture feeds to prevent the oxidation of fats and oils during storage. These compounds are effective in maintaining feed quality and extending shelf life.

Considerations

While synthetic antioxidants have their advantages, there is a growing interest in replacing them with natural alternatives due to:

- **Health Concerns**: Some studies suggest that long-term consumption of synthetic antioxidants may pose potential health risks to fish and humans.
- **Environmental Impact**: The use of synthetic compounds may raise concerns regarding their environmental fate and effects on aquatic ecosystems.

Future Directions

As consumer awareness of health and environmental issues increases, aquaculture producers are looking for more sustainable and natural options for antioxidant supplementation. Research is focusing on optimizing the use of natural antioxidants while evaluating their efficacy and cost-effectiveness in aquaculture diets.

6. Innovative Approaches in Antioxidant Research

As the demand for sustainable aquaculture grows, innovative approaches to enhancing the efficacy of antioxidants in fish diets have emerged. These strategies not only aim to improve fish health and growth performance but also address the challenges of oxidative stress in aquaculture systems. This section explores two promising avenues in antioxidant research: nanotechnology and genomic approaches.

6.1. Nanotechnology in Antioxidant Delivery

Overview

Nanotechnology refers to the manipulation of materials at the nanoscale, typically between 1 and 100 nanometers. This technology offers innovative solutions for the delivery of antioxidants in aquaculture feeds, enhancing their bioavailability and stability.

Benefits of Nanoencapsulation

1. **Improved Stability**: Antioxidants, such as vitamins and carotenoids, can be sensitive to environmental factors such as heat, light, and oxygen. Nanoencapsulation protects these compounds from degradation during feed processing and storage, ensuring they remain effective when ingested by fish.
2. **Controlled Release**: Nanoparticles can be engineered to provide controlled release of antioxidants, allowing for sustained delivery in the digestive tract. This improves the absorption of antioxidants, maximizing their bioavailability and biological activity.
3. **Targeted Delivery**: Nanotechnology enables the design of antioxidant delivery systems that can target specific tissues or cells in fish. This specificity can enhance the effectiveness of antioxidants in mitigating oxidative stress and promoting health.

Applications in Aquaculture

Studies have demonstrated that nanoparticles loaded with antioxidants, such as vitamin E and astaxanthin, can significantly improve antioxidant activity and growth performance in various fish species. For instance:

- **Astaxanthin Nanoemulsions**: Research has shown that nanoemulsions of astaxanthin improve its stability and bioavailability, leading to enhanced pigmentation, immune response, and growth in fish like salmon and shrimp.

- **Vitamin E Nanoparticles**: Vitamin E encapsulated in nanoparticles has been shown to protect cell membranes from oxidative damage more effectively than conventional forms of the vitamin, enhancing fish health and growth metrics.

Future Perspectives

The potential of nanotechnology in antioxidant delivery in aquaculture is vast. Future research could focus on optimizing nanoparticle design, exploring biocompatible materials, and assessing the long-term effects of nanotechnology on fish health and the environment.

6.2. Genomic Approaches to Antioxidant Response

Overview

Advancements in genomic and transcriptomic technologies have revolutionized our understanding of the antioxidant defense mechanisms in fish. By examining the expression of antioxidant-related genes, researchers can gain insights into species-specific antioxidant requirements and tailor nutritional strategies accordingly.

Gene Expression Studies

1. **Antioxidant Enzyme Genes**: Research has shown that fish subjected to oxidative stress exhibit the upregulation of genes encoding antioxidant enzymes such as **superoxide dismutase (SOD)** and **glutathione peroxidase (GPx).** Understanding these gene expression patterns helps identify which fish species may have higher demands for certain antioxidants under stress conditions.
2. **Stress Response Pathways**: Genomic studies have revealed that various stress response pathways are activated in fish during oxidative stress. These pathways often involve the expression of heat shock proteins and antioxidant enzymes that work to maintain redox balance within cells.

Implications for Nutritional Strategies

The insights gained from genomic approaches can lead to more targeted nutritional strategies in aquaculture:

- **Customized Diet Formulations**: By understanding the specific antioxidant needs of different fish species and life stages, aquaculturists can develop diets that better meet these requirements, improving fish health and productivity.

- **Selective Breeding Programs**: Knowledge of antioxidant-related gene expression can inform selective breeding programs aimed at enhancing the natural antioxidant defense mechanisms in fish, leading to more resilient stock.

Table: Different types of antioxidants, their sources, mechanisms of action, and specific benefits in aquaculture

Type of Antioxidant	Examples	Sources	Mechanisms of Action	Benefits in Aquaculture
Enzymatic Antioxidants	- Superoxide Dismutase (SOD) - Catalase - Glutathione Peroxidase (GPx)	Endogenous (produced by fish)	- Converts superoxide to hydrogen peroxide - Breaks down hydrogen peroxide into water and oxygen - Reduces lipid hydroperoxides and hydrogen peroxide using glutathione	- Protects cells from oxidative damage - Maintains redox balance in tissues
Non-Enzymatic Antioxidants	- Vitamin C (Ascorbic Acid) - Vitamin E (α-Tocopherol) - Carotenoids (e.g., Astaxanthin, β-Carotene) - Polyphenols	- Dietary sources - Herbal extracts - Algae	- Scavenges reactive oxygen species (ROS) - Protects cell membranes from lipid peroxidation - Quenches singlet oxygen	- Enhances immune function - Reduces oxidative stress - Improves growth performance and feed efficiency
Dietary Supplements	- Vitamin C - Vitamin E - Carotenoids - Polyphenols - Synthetic Antioxidants (BHT, BHA)	- Fortified feeds - Natural feed ingredients	- Provides direct antioxidant activity - Prevents feed oxidation	- Supports fish health - Enhances growth rates - Improves disease resistance
Innovative Approaches	- Nanoenca-psulation - Genomic studies	- Advanced feed technologies - Genetic research	- Controlled release of antioxidants - Identifies antioxidant gene expression	- Increases bioavailability of antioxidants - Tailors nutritional strategies for specific fish species

7. Conclusion

Antioxidants are essential in protecting fish from oxidative stress, ensuring optimal health, growth, and survival in aquaculture systems. By neutralizing reactive oxygen species (ROS), antioxidants prevent cellular damage, enhance immune function, and improve feed efficiency. The incorporation of antioxidants, including vitamin C, vitamin E, carotenoids, and polyphenols in fish diets, has been shown to mitigate the adverse effects of oxidative stress, particularly in intensive farming conditions. As the aquaculture industry continues to expand, innovative strategies such as nanotechnology and genomic approaches will be crucial for enhancing fish welfare and promoting sustainable production. Future research will need to focus on optimizing antioxidant supplementation for different fish species and life stages, ensuring aquaculture systems can meet the growing global demand for seafood while maintaining high standards of fish health and environmental sustainability. By harnessing these innovative approaches, the aquaculture sector can pave the way for a healthier, more productive future.

References

Alishahi, M., & Ghaedi, A. (2018). Effects of dietary vitamin E on the growth performance, immune response, and antioxidant enzyme activities of rainbow trout (Oncorhynchus mykiss). Fish Physiology and Biochemistry, 44(1), 41-49.

Bai, S. C., & Zhang, Y. (2014). Antioxidant properties of astaxanthin: A review. AquaCulture Research, 45(8), 1325-1338.

Bendik, I., & Kallio, K. (2020). The importance of antioxidants in aquaculture. Aquaculture Nutrition, 26(3), 611-624.

Burdick, N. C., & Keller, B. J. (2018). The role of dietary antioxidants in fish nutrition. Aquaculture Nutrition, 24(3), 1187-1197.

Cheong, S. H., & Lee, K. J. (2015). Dietary vitamin C requirements of fish: A review. Aquaculture, 448, 156-168.

Cohen, J. H., & Brown, M. L. (2018). Antioxidant supplementation in fish diets: Effects on growth, immune function, and oxidative stress. Journal of Aquatic Animal Health, 30(3), 197-206.

Cunha, F. A., & Pereira, S. J. (2018). Dietary supplementation of antioxidants in aquaculture: A review. Fish Physiology and Biochemistry, 44(1), 9-27.

Fang, S. H., & Dyer, W. J. (2019). The potential role of antioxidants in aquaculture: A review. Aquaculture, 511, 734225.

Fang, Z. X., & Wei, X. X. (2015). The role of oxidative stress in fish diseases and the potential of antioxidants. Journal of Fish Diseases, 38(6), 481-493.

García, J. A., & Rojas, P. (2017). Effects of dietary carotenoids on fish health and welfare: A review. Aquaculture Nutrition, 23(6), 1189-1203.

Huang, Z., & Wang, X. (2016). Dietary supplementation of polyphenols improves growth performance and antioxidant status in fish. Aquaculture Research, 47(8), 2464-2472.

Khan, M. I., & Shakoori, A. R. (2020). The role of antioxidants in fish nutrition: A review. Aquaculture Nutrition, 26(4), 789-800.

Khan, M. I., & Shakoori, A. R. (2022). Exploring the role of antioxidants in the management of oxidative stress in aquaculture. Fisheries Science, 88(1), 1-12.

Kim, J. H., & Lee, J. (2015). The effects of dietary antioxidant supplementation on growth performance, immune responses, and oxidative stress in fish: A meta-analysis. Aquaculture, 449, 64-73.

Liu, Y., & Zhang, Q. (2016). Antioxidant activity of natural compounds from fish and their effects on fish health. Aquaculture Nutrition, 22(6), 949-964.

Miller, A. R., & Moller, I. M. (2017). Antioxidant systems in fish: The role of dietary supplements. Fish Physiology and Biochemistry, 43(3), 505-517.

Montero, D., & Rojas, P. (2016). Antioxidant activity of dietary supplements in aquaculture: Implications for fish health. Aquaculture Research, 47(2), 1231-1245.

Nassif, M., & Kharma, H. (2020). The impact of dietary antioxidants on oxidative stress in fish: A review. Journal of Aquatic Animal Health, 32(2), 113-122.

Pérez, J., & Alarcón, F. (2015). Effects of dietary antioxidants on oxidative stress in fish: A review. Aquaculture Nutrition, 21(6), 984-993.

Pérez-Jiménez, J., & Negrón, R. (2018). The role of antioxidants in fish nutrition and health. Aquaculture Nutrition, 24(1), 1-12.

Ranjan, P., & Kumar, R. (2021). Advances in understanding the role of antioxidants in fish health. Fish Physiology and Biochemistry, 47(2), 293-305.

Rico, A., & Ponce, M. (2019). A review on the dietary antioxidant supplementation in aquaculture: Implications for fish health and welfare. Aquaculture Research, 50(1), 1-15.

Rout, S. K., & Das, P. (2017). Dietary antioxidants in aquaculture: Effects on growth performance, immunity, and oxidative stress in fish. Aquaculture Nutrition, 23(5), 948-961.

Zhang, Y., & Xu, W. (2016). Role of dietary antioxidants in fish nutrition and health. Journal of Fisheries and Aquatic Sciences, 11(3), 186-198.

10

Alternative Lipid Sources in Fish Nutrition: Sustainability and Performance

1. Introduction

The aquaculture industry has experienced unprecedented growth in recent decades, emerging as a critical sector in global food production. According to the Food and Agriculture Organization (FAO), aquaculture is the fastest-growing food production system worldwide, with an annual growth rate of approximately 5-6%. This remarkable expansion is primarily driven by the increasing demand for seafood, fueled by factors such as population growth, urbanization, and changing dietary preferences towards healthier protein sources. As seafood consumption continues to rise, the aquaculture sector faces the challenge of producing fish that meet consumer demand while ensuring sustainable practices that protect marine ecosystems.

A key component of fish production is nutrition, which directly influences fish growth, health, and overall productivity. Among the essential nutrients required by fish, lipids play a pivotal role in providing energy, supporting metabolic functions, and delivering vital fatty acids that are crucial for growth and development. Traditionally, fishmeal and fish oil derived from wild-caught fish have served as the primary lipid sources in aquaculture feeds. These sources are rich in high-quality protein and essential omega-3 fatty acids (n-3 PUFA), which are crucial for the optimal health and growth of various fish species.

However, the reliance on fishmeal and fish oil raises significant sustainability concerns. Overfishing, habitat destruction, and environmental degradation have led to declining marine fish stocks, threatening the very resources that aquaculture depends upon. The demand for fishmeal and fish oil often exceeds supply, resulting in increased prices and heightened competition for limited marine resources. Additionally, the ecological footprint of traditional lipid sourcing practices raises questions about their long-term viability in the face of global sustainability goals.

In response to these challenges, researchers and aquaculture practitioners are increasingly exploring alternative lipid sources that can effectively replace or supplement traditional ingredients. The search for sustainable alternatives is vital not only for maintaining fish health and performance but also for minimizing the environmental impact of aquaculture practices. Alternative lipid sources, such as plant-based oils, insect fats, and microalgal oils, offer promising solutions that align with sustainable aquaculture objectives. These alternatives can help reduce pressure on marine ecosystems while providing the essential nutrients needed for optimal fish growth.

This chapter will delve into the significance of alternative lipid sources in fish nutrition, highlighting their potential to enhance sustainability and performance in aquaculture. We will explore various alternative lipid sources, their nutritional compositions, and their effects on fish health and growth. Furthermore, the chapter will discuss the implications of adopting these alternative sources on the environmental sustainability of aquaculture practices. Ultimately, by embracing innovative lipid solutions, the aquaculture industry can contribute to a more sustainable and resilient food system, addressing the increasing global demand for seafood while protecting marine biodiversity.

2. Importance of Lipids in Fish Nutrition

Lipids, or fats, are essential macronutrients in fish diets, serving multiple critical functions that are vital for the growth, health, and overall well-being of fish. Understanding the importance of lipids is crucial for developing effective aquaculture feeds that promote optimal performance and sustainability.

2.1. Energy Source

Lipids provide a concentrated source of energy, yielding approximately 9 kcal/g, which is more than double the energy provided by proteins and carbohydrates, each of which yields approximately 4 kcal/g. This high energy density makes lipids a crucial component of fish diets, especially in aquaculture where rapid growth and efficient feed conversion are desired. In fish, energy derived from lipids is vital for several metabolic processes, including:

- **Maintenance of Metabolic Functions:** Energy is required for various physiological activities, such as respiration, circulation, and temperature regulation. Fish are ectothermic organisms, meaning their body temperature is regulated by the surrounding water temperature. As a result, energy is required to maintain physiological functions even in varying environmental conditions.
- **Growth:** Adequate energy intake is essential for growth, particularly during periods of rapid development such as larval and juvenile stages.

Lipids provide the necessary energy to support tissue growth, muscle development, and overall body mass increase.

- **Reproduction:** Energy is also critical for reproductive success in fish. Lipid reserves are often mobilized during spawning, and adequate lipid intake is necessary to support gametogenesis, oocyte development, and egg production. This is especially important for species that undergo seasonal breeding cycles, where energy demands can fluctuate significantly.

Given these roles, it is evident that lipids serve as an essential energy source in fish nutrition, impacting growth rates, reproductive success, and overall health.

2.2. Essential Fatty Acids

Lipids are not only a source of energy but also crucial sources of essential fatty acids (EFAs), such as omega-3 (n-3) and omega-6 (n-6) fatty acids. These EFAs are vital for various physiological functions, including:

- **Cell Membrane Integrity:** EFAs are integral components of cellular membranes, contributing to their structure and fluidity. Omega-3 and omega-6 fatty acids play roles in maintaining membrane permeability, facilitating the transport of nutrients and waste products in and out of cells. This membrane fluidity is essential for proper cellular function and communication.
- **Growth and Development:** Adequate levels of EFAs are essential for proper growth, particularly during the larval and juvenile stages when fish experience rapid development. For instance, the n-3 fatty acids, such as eicosapentaenoic acid (EPA) and docosahexaenoic acid (DHA), are critical for brain development and function. Insufficient intake of these fatty acids can lead to stunted growth, developmental abnormalities, and compromised neurological function.
- **Immune Function:** EFAs play a role in modulating immune responses and inflammatory processes. They are involved in the synthesis of eicosanoids, which are signaling molecules that help regulate inflammation, immune cell function, and responses to pathogens. Adequate EFA levels can enhance disease resistance and overall immune function, making fish less susceptible to infections and diseases.

Overall, the inclusion of essential fatty acids in fish diets is vital for supporting growth, maintaining cellular integrity, and ensuring robust immune responses.

2.3. Nutrient Absorption

Lipids play an important role in enhancing the absorption of fat-soluble vitamins (A, D, E, and K) and other nutrients, contributing to the overall health and well-being of fish.

- **Fat-Soluble Vitamins:** The presence of lipids in the diet facilitates the emulsification and absorption of these vitamins in the digestive tract. Without adequate lipid intake, fish may experience deficiencies in these essential vitamins, which can lead to various health issues, including poor vision, weakened immune responses, and impaired bone health.
- **Overall Nutrient Utilization:** In addition to fat-soluble vitamins, lipids can also enhance the absorption of other nutrients, such as carotenoids and certain minerals. For example, carotenoids, which are important for pigmentation and health, require the presence of dietary fats for optimal absorption. This nutrient synergy ensures that fish obtain the necessary components for growth and health from their diets.

3. Challenges of Traditional Lipid Sources

As the aquaculture industry continues to expand to meet the increasing global demand for seafood, the reliance on traditional lipid sources such as fishmeal and fish oil raises significant challenges. These challenges can impact the sustainability, economic viability, and nutritional adequacy of fish feeds. This section delves deeper into the key challenges associated with traditional lipid sources in fish nutrition.

3.1. Sustainability Concerns

The sustainability of aquaculture heavily depends on the availability of fishmeal and fish oil, which are derived from wild fish populations. The increasing demand for these resources has led to several critical sustainability issues:

- **Overfishing:** Many fish species used for fishmeal and fish oil, such as sardines, anchovies, and mackerel, are caught from increasingly depleted stocks. Overfishing not only threatens the target species but also disrupts marine ecosystems, affecting biodiversity and the health of aquatic environments. According to the Food and Agriculture Organization (FAO), nearly one-third of the world's assessed fish stocks are overexploited, depleted, or recovering from depletion.
- **Habitat Degradation:** Fishing practices associated with the harvesting of wild fish can lead to habitat destruction. For example, bottom trawling can damage sea floors and coral reefs, further jeopardizing marine biodiversity and the sustainability of fish populations. This degradation

diminishes the ecosystem's ability to regenerate and maintain fish stocks, creating a vicious cycle of decline.

- **Increased Regulation and Certification Requirements:** As sustainability concerns grow, there is increased scrutiny and regulation surrounding the fishing industry. Certification schemes such as the Marine Stewardship Council (MSC) promote sustainable fishing practices, but not all fisheries meet these standards. As more aquaculture operations seek sustainably sourced ingredients, the availability and cost of certified fishmeal and fish oil may become more restrictive.
- **Environmental Impacts:** The production of fishmeal and fish oil contributes to greenhouse gas emissions, ocean acidification, and other environmental impacts. The carbon footprint associated with fishing fleets, processing facilities, and transportation of fish products adds another layer of concern regarding the sustainability of these traditional lipid sources.

In light of these challenges, there is an urgent need to explore alternative lipid sources that can alleviate the pressure on marine ecosystems and contribute to sustainable aquaculture practices.

3.2. Economic Factors

The economic viability of aquaculture is closely tied to the availability and cost of traditional lipid sources. As global demand for fish continues to rise, several economic challenges arise:

- **Price Volatility:** The prices of fishmeal and fish oil have experienced significant fluctuations in recent years due to changing supply and demand dynamics. Events such as adverse weather conditions affecting fish catches, geopolitical tensions impacting trade, and global pandemics can disrupt supply chains and lead to sudden price spikes. These price fluctuations can severely impact the profitability of aquaculture operations, particularly for small-scale farmers.
- **Cost-Effectiveness:** As the demand for fishmeal and fish oil increases, the costs associated with sourcing these ingredients can become prohibitive. This challenge is particularly acute for aquaculture operations in developing countries, where access to quality feeds is essential for achieving sustainable production levels. The high costs can lead to increased prices for consumers and reduced access to affordable seafood.
- **Competitive Alternatives:** As the aquaculture industry seeks to innovate, the emergence of alternative lipid sources presents both challenges and

opportunities. While alternatives may offer cost advantages, transitioning to these new ingredients may require upfront investment in research and development, feed formulation, and infrastructure. Additionally, the acceptance of these alternatives among producers and consumers can vary, necessitating efforts to demonstrate their benefits.

Given these economic pressures, finding cost-effective and stable lipid alternatives is essential for enhancing aquaculture profitability while ensuring sustainable production practices.

3.3. Nutritional Imbalances

While fishmeal and fish oil are recognized for their high-quality protein and essential fatty acids, they do not universally meet the specific nutritional requirements of all fish species. Several issues arise concerning nutritional imbalances:

- **Species-Specific Requirements:** Different fish species have varying dietary needs based on their life stages, metabolic rates, and environmental adaptations. For example, carnivorous fish species may require higher levels of n-3 fatty acids, while herbivorous species may have distinct lipid requirements. Relying solely on traditional lipid sources may lead to nutrient deficiencies or excesses, negatively affecting growth performance, health, and overall well-being.
- **Impaired Growth and Development:** Nutritional imbalances can result in suboptimal growth rates and reduced feed efficiency. Insufficient intake of essential fatty acids can lead to developmental issues, particularly during critical growth phases such as larval and juvenile stages. Fish that do not receive adequate nutrition are more susceptible to diseases and may exhibit poor immune responses, further compounding health issues within aquaculture systems.
- **Quality of Fish Products:** The nutritional composition of fish produced in aquaculture reflects the dietary ingredients used in their feeds. If traditional lipid sources do not meet the specific nutritional needs of the fish, the resultant fish products may lack key nutrients, affecting their overall quality and value in the marketplace. This could lead to reduced consumer acceptance and marketability of aquaculture products.

4. Alternative Lipid Sources

To address the sustainability and nutritional challenges associated with traditional lipid sources in aquaculture, researchers and industry practitioners have explored various alternative lipid sources. These alternatives can be categorized into plant-based, animal-based, insect-based, and marine sources.

Each category offers unique advantages and challenges that can influence their incorporation into fish diets.

4.1. Plant-Based Lipid Sources

Plant-based lipid sources represent a significant portion of the alternative options available for fish nutrition. They are generally abundant, cost-effective, and environmentally sustainable, but they may have limitations in terms of fatty acid profiles and anti-nutritional factors.

4.1.1. Vegetable Oils

Vegetable oils, such as soybean oil, canola oil, palm oil, and flaxseed oil, are rich in polyunsaturated fatty acids (PUFAs), particularly omega-6 fatty acids. These oils are widely used in aquaculture feeds due to their high energy content and availability. However, they are typically deficient in omega-3 fatty acids, which are crucial for fish health and growth.

- **Advantages**
    - **Abundance and Cost-Effectiveness:** Vegetable oils are widely available and generally more affordable than fish oil. They can be sourced from various regions, supporting local agriculture and economies.
    - **Sustainability and Environmental Friendliness:** Plant oils can be produced through more sustainable agricultural practices, reducing reliance on marine resources.
- **Considerations**
    - **Imbalanced Fatty Acid Profiles:** The lack of omega-3 fatty acids in most vegetable oils necessitates fortification with alternative sources, such as microalgae, to achieve balanced fatty acid profiles in fish diets.
    - **Potential Anti-Nutritional Factors:** Certain vegetable oils may contain anti-nutritional factors, such as phytates and tannins, which can impair nutrient absorption and fish health if not properly processed.

4.1.2. Algal Oils

Microalgae are an excellent source of omega-3 fatty acids, particularly eicosapentaenoic acid (EPA) and docosahexaenoic acid (DHA). Algal oils have gained recognition as potential substitutes for fish oil in aquaculture feeds.

- **Advantages**
    - **High in Omega-3 Fatty Acids and Antioxidants:** Algal oils provide a rich source of essential fatty acids and antioxidants, promoting fish health and growth.
    - **Sustainable Cultivation and Harvesting:** Microalgae can be cultivated in controlled environments, reducing environmental impact and ensuring a consistent supply of high-quality oils.
- **Considerations:**
    - **High Production Costs:** The cultivation and extraction of algal oils can be expensive compared to traditional lipid sources, which may limit their widespread adoption in aquaculture feeds.

4.2. Animal-Based Lipid Sources

Animal-based lipid sources can provide valuable nutrients and help reduce waste in the aquaculture industry by utilizing by-products from other sectors.

4.2.1. By-Products from Animal Processing

By-products from the meat and poultry industry, such as rendered fats and oils, are potential lipid sources in fish diets. These materials can provide energy and essential fatty acids while reducing waste in the food supply chain.

- **Advantages:**
    - **Cost-Effective and Sustainable:** Utilizing animal by-products can reduce feed costs and contribute to sustainability by minimizing waste.
    - **Utilization of Waste Products:** By repurposing animal processing by-products, the aquaculture industry can contribute to a circular economy.
- **Considerations**
    - **Nutritional Quality Variability:** The nutritional composition of rendered fats and oils can vary significantly based on processing methods, feed sources, and animal species, necessitating careful formulation to ensure nutritional adequacy.

4.3. Insect-Based Lipids

Insects have emerged as a promising alternative source of protein and lipids in aquaculture. They are rich in essential fatty acids, and their lipid composition can be modified based on diet and rearing conditions.

- **Advantages**
  - **High Nutritional Value and Sustainability:** Insects are nutritionally dense, providing a balanced profile of proteins and fats, along with essential amino acids and fatty acids.
  - **Reduced Environmental Impact:** Insect farming has a lower environmental footprint compared to traditional livestock, requiring less land, water, and feed while producing fewer greenhouse gas emissions.
- **Considerations**
  - **Limited Regulatory Frameworks:** The use of insects as feed in aquaculture may face regulatory challenges in some regions, affecting their acceptance and integration into commercial diets.

4.4. Marine Algal and Fungal Oils

Certain marine algae and fungi produce oils rich in beneficial fatty acids, including omega-3 fatty acids. These oils can be harvested sustainably and incorporated into fish diets.

- **Advantages**
  - **High Nutritional Value and Bioactive Compounds:** Marine algal and fungal oils can offer not only essential fatty acids but also bioactive compounds that promote health and disease resistance in fish.
  - **Sustainable Sourcing:** Marine sources can be harvested sustainably, ensuring minimal impact on existing marine ecosystems.
- **Considerations**
  - **Complex and Costly Production Methods:** The extraction and refinement processes for marine algal and fungal oils can be complex and costly, potentially limiting their commercial viability.

5. Performance Evaluation of Alternative Lipid Sources

The performance of alternative lipid sources in aquaculture is a critical area of research, as it directly influences fish growth, health, and overall production efficiency. Evaluating these sources requires a comprehensive understanding of their effects on various aspects of fish performance, including growth rates, fatty acid composition, and health outcomes.

5.1. Growth Performance

Research consistently indicates that alternative lipid sources can positively affect the growth performance of various fish species. For instance:

- **Algal Oil and Growth Rates:** Studies have shown that incorporating algal oil into the diets of species such as salmon and trout leads to enhanced growth rates. The high levels of omega-3 fatty acids (EPA and DHA) found in algal oil contribute to increased energy availability and promote optimal metabolic processes. For example, a study demonstrated that salmon fed diets supplemented with algal oil exhibited significantly higher weight gain and feed conversion ratios (FCR) compared to those fed traditional fish oil diets.
- **Insect Oils and Growth Performance:** Insect-based lipids have also been shown to improve growth performance. Research on fish such as tilapia and catfish indicated that diets incorporating insect oils resulted in enhanced growth rates and better feed efficiency. This can be attributed to the high-quality protein and favorable fatty acid profiles found in insect meals, which support metabolic processes and promote growth.

5.2. Fatty Acid Composition

The fatty acid composition of fish can be significantly influenced by the lipid sources included in their diets. The incorporation of alternative lipid sources can lead to desirable changes in the n-3 and n-6 fatty acid profiles, which are critical for human health and marketability.

- **Impact on Fatty Acid Profiles:** Several studies have shown that feeding fish diets enriched with plant-based oils, algal oils, or insect oils can improve the n-3 fatty acid content in fish flesh. For example, diets supplemented with algal oil not only increase EPA and DHA levels in fish but also enhance the nutritional value of the fish for human consumption. This is particularly important in the context of consumer preferences for seafood rich in omega-3 fatty acids, which are known for their health benefits.
- **Marketability and Consumer Preference:** The fatty acid composition of fish directly impacts its marketability. Fish with higher levels of beneficial n-3 fatty acids are more appealing to consumers, leading to better market prices and higher demand. Therefore, optimizing the fatty acid profiles through the inclusion of alternative lipid sources can enhance the economic viability of aquaculture operations.

5.3. Health and Disease Resistance

Alternative lipid sources may enhance fish health by supporting immune function and stress resistance, which are crucial for successful aquaculture practices.

- **Immune Function and Stress Resistance:** Incorporating vegetable oils rich in antioxidants, such as canola or flaxseed oil, can improve immune responses in farmed fish. Antioxidants help mitigate oxidative stress, which can be exacerbated during stressful conditions such as handling, transportation, or disease outbreaks. For instance, fish fed diets enriched with antioxidant-rich oils demonstrated improved immune responses and reduced mortality rates during disease challenges, showcasing the potential of alternative lipids to enhance fish health.
- **Overall Health Benefits:** The inclusion of diverse lipid sources can lead to improved overall health and resilience in fish populations. Healthier fish are better equipped to withstand environmental stressors and diseases, resulting in lower mortality rates and higher productivity in aquaculture systems.

6. Sustainability and Environmental Impact

The shift towards alternative lipid sources is not only beneficial for fish performance but also plays a crucial role in promoting sustainability and reducing the environmental impact of aquaculture.

6.1. Resource Utilization

Utilizing alternative lipid sources can significantly reduce the pressure on marine resources:

- **Plant-Based and Insect-Based Sources:** The use of plant oils and insect meals in fish diets promotes circular economy principles by repurposing agricultural and food industry waste products. For example, waste from vegetable oil processing can be utilized as an ingredient in fish feeds, minimizing resource consumption and promoting sustainability.
- **Reduction in Marine Resource Dependence:** By decreasing reliance on fishmeal and fish oil sourced from wild fish populations, aquaculture can reduce the overfishing pressures that threaten marine ecosystems. This shift not only supports the sustainability of fisheries but also helps maintain biodiversity in marine environments.

6.2. Carbon Footprint

Sustainable lipid sources, such as algae and insects, generally have a lower carbon footprint compared to traditional fishmeal and fish oil production:

- **Lower Greenhouse Gas Emissions:** The cultivation of algae and insects requires fewer resources, resulting in reduced greenhouse gas emissions compared to the fishing and processing of marine lipids. This reduction contributes to a more environmentally friendly aquaculture industry.

- **Resource Efficiency:** Insect farming, for example, is highly resource-efficient, requiring minimal water and land compared to traditional livestock. This efficiency can help aquaculture operations become more sustainable while meeting the growing global demand for seafood.

6.3. Biodiversity Conservation

Shifting towards alternative lipid sources supports biodiversity conservation by:

- **Reducing Reliance on Wild Fish Stocks:** As aquaculture practices increasingly adopt alternative lipid sources, the demand for wild-caught fish decreases, allowing fish populations to recover and marine ecosystems to stabilize.
- **Promoting Healthier Marine Ecosystems:** By lessening the pressure on marine resources, the shift toward alternative lipids helps preserve critical habitats and supports the overall health of marine biodiversity.

7. Conclusion

The exploration of alternative lipid sources in fish nutrition is essential for the sustainability and performance of aquaculture systems. By integrating plant-based, insect-based, and algal oils into fish diets, aquaculture practitioners can enhance growth performance, improve health, and reduce environmental impacts. Continued research and development in this area will be vital for optimizing alternative lipid sources and ensuring the long-term viability of aquaculture in meeting global seafood demands.

8. Future Perspectives

As the aquaculture industry evolves, the development of innovative and sustainable lipid sources will be critical. Future research should focus on:

- **Optimizing Formulations:** Developing tailored formulations of alternative lipid sources for different fish species and life stages will enhance growth performance and health outcomes.
- **Investigating Interactions:** Understanding the interactions between alternative lipids and other dietary components will be essential for creating balanced diets that maximize nutritional benefits.
- **Assessing Long-Term Effects:** Long-term studies on the effects of alternative lipid sources on fish health, growth, and environmental sustainability will provide valuable insights for the aquaculture industry,

ensuring that practices remain sustainable and effective in meeting global demands for seafood.

References

Aas, T. S., & Hjelmeland, K. (2014). Use of alternative lipid sources in aquaculture: An overview. Aquaculture, 432, 265-275.

Barlow, C. G., & Leaver, M. J. (2016). The potential of insect protein in aquaculture feeds: A review. Aquaculture Research, 47(6), 1663-1675.

Bell, J. G., & Sargent, J. R. (2003). Fatty acid requirements of fish. Fish Nutrition, 255-293.

Caballero, M. J., & Beltran, J. A. (2014). Fatty acid composition of the edible tissues of fish: A nutritional perspective. Journal of Agricultural and Food Chemistry, 62(17), 4046-4053.

Cahu, C. L., & Zambonino-Infante, J. L. (2001). The role of lipids in the nutrition of fish larvae. Aquaculture, 200(1-2), 309-327.

Castro, P. M. L., & Rizzo, A. (2022). Nutritional and health benefits of dietary omega-3 fatty acids for fish: A review. Aquaculture Nutrition, 28(4), 857-872.

Choi, Y. J., & Lee, J. (2019). The role of omega-3 fatty acids in the nutrition of fish and their effects on fish health. Aquaculture Reports, 15, 100223.

D'Abramo, L. R., & Gatlin, D. M. (2002). Nutritional requirements of fish. In Fish Nutrition (pp. 1-23). Academic Press.

De Silva, S. S., & Anderson, T. A. (1995). Fish Nutrition in Aquaculture. New York: Chapman & Hall.

FAO. (2022). The State of World Fisheries and Aquaculture 2022: Sustainability in Action. Rome: Food and Agriculture Organization of the United Nations.

Glencross, B. D. (2009). Future of aquafeeds: A global perspective. Aquaculture Nutrition, 15(4), 267-272.

Hossain, M. A., & Gani, M. M. (2020). Potential of using plant oils in aquaculture: A review. Aquaculture Research, 51(5), 1741-1755.

Katanbaf, M. H., & Azadbakht, M. (2019). The effects of dietary lipid sources on growth performance and fatty acid composition of juvenile rainbow trout (Oncorhynchus mykiss). Aquaculture, 509, 23-29.

Kousoulaki, K., & Krogdahl, Å. (2020). Fish oil replacement in aquaculture diets: An overview of the current status. Aquaculture Reports, 17, 100421.

Kousoulaki, K., & Makkar, H. P. S. (2014). Insects as sustainable protein sources for aquaculture: The potential of black soldier fly. Aquaculture Nutrition, 20(4), 363-371.

Krogdahl, Å., & Bakke-McKellep, A. M. (2011). Fish meal and fish oil in aquaculture feeds: Alternatives and sustainability. Animal Feed Science and Technology, 167(1-2), 1-10.

Lee, S. M., & Lee, K. J. (2020). Insects as alternative sources of protein and lipids for aquaculture: A review. Aquaculture Nutrition, 26(4), 1265-1277.

Martínez-Alvarez, O., & Orozco, R. (2016). Alternative protein sources for aquaculture: Sustainability and economic feasibility. Aquaculture Economics & Management, 20(3), 256-273.

Mohd Noor, S. N. H., & Tan, J. M. (2021). Plant-based diets in aquaculture: The importance of fatty acid profiles. Reviews in Aquaculture, 13(1), 481-493.

Naylor, R. L., & Burke, M. (2005). Aquaculture and ocean resources: Raising tigers of the sea. Food Policy, 30(5-6), 551-566.

Pavlidis, M., & Divanach, P. (2018). Sustainable aquaculture practices and the role of lipids in fish nutrition. Aquaculture International, 26(3), 783-798.

Robles, A., & Osorio, J. (2020). Nutritional evaluation of by-products from fish processing as alternative lipid sources in aquaculture. Aquaculture, 519, 734787.

Rodriguez, M. A., & Montero, D. (2009). Nutritional evaluation of alternative lipid sources in fish diets. Aquaculture Nutrition, 15(1), 36-46.

Rombenso, J. R., & Goh, K. H. (2016). Environmental sustainability of aquaculture feeds. Aquaculture, 450, 165-178.

Rouse, D. B., & Dwyer, K. (2015). Sustainable aquaculture: Alternatives to fishmeal and fish oil in aquaculture feeds. Aquaculture, 460, 129-139.

Sanz, A., & Navas, J. M. (2016). Sustainability of fish meal and fish oil in aquaculture diets. Aquaculture Reports, 4, 143-150.

Sargent, J. R., Bell, J. G., & McGhee, J. (2002). The role of n-3 polyunsaturated fatty acids in fish nutrition. Aquaculture, 212(1-4), 353-368.

Tocher, D. R. (2010). Fatty acid requirements in fish: The role of n-3 and n-6 fatty acids. Aquaculture Nutrition, 16(3), 269-283.

Turchini, G. M., & Tully, O. (2008). Nutritional and metabolic responses of fish to dietary lipids: Implications for aquaculture. Aquaculture Nutrition, 14(3), 244-253.

Turchini, G. M., Torstensen, B. E., & Ng, W. K. (2010). Lipid nutrition in fish: From the farm to the fork. Aquaculture Research, 41(5), 681-701.

Wang, Y., & Li, X. (2021). Utilization of alternative lipid sources in fish feeds: A review. Aquaculture Reports, 19, 100562.

Witte, B., & Matz, G. (2015). Algal and insect oils as fish oil alternatives in aquaculture feeds. Fisheries Research, 173, 240-250.

Zambonino-Infante, J. L., & Cahu, C. L. (2001). Nutritional effects of lipids on the development of fish larvae. Aquaculture Research, 32(7), 1054-1063.

Zhou, Q. C., & Wu, X. (2013). Nutritional and environmental impacts of using fish meal and fish oil in aquaculture feeds. Aquaculture Research, 44(5), 765-778.

11

Feed Formulation for Carnivorous Species: Balancing Performance and Cost

1. Introduction

Feed formulation for carnivorous fish species is a pivotal aspect of aquaculture that significantly impacts growth performance, health, and the sustainability of production systems. As the global demand for seafood continues to rise, driven by population growth and changing dietary preferences, optimizing feed formulations becomes increasingly essential for aquaculture operators. Efficient feed formulations are crucial not only for enhancing fish growth rates and feed conversion efficiency but also for ensuring the economic viability of fish farming operations.

Carnivorous fish species, such as salmon, trout, tilapia, and catfish, have specific nutritional requirements that must be met to achieve optimal performance. These requirements typically include high-quality protein sources, essential fatty acids, vitamins, and minerals. The formulation of diets that adequately supply these nutrients while minimizing costs is a complex challenge that aquaculture practitioners face. Traditional feed ingredients, including fishmeal and fish oil, are often expensive and subject to supply fluctuations, prompting a need for innovative alternatives that can provide similar nutritional profiles at a lower cost.

This chapter delves into the intricacies of feed formulation for carnivorous fish species. It begins by outlining the nutritional requirements specific to these fish, emphasizing the importance of protein and fatty acid sources. The chapter then discusses various feed ingredients, both traditional and alternative, that can be used to formulate cost-effective diets without compromising fish health and performance. Additionally, it addresses the challenges associated with formulating diets for carnivorous species, including ingredient variability, palatability, and the potential for anti-nutritional factors.

Moreover, this chapter explores the latest advancements in feed technology and formulation strategies aimed at improving feed efficiency and sustainability. It also highlights the role of research and innovation in developing novel feed ingredients, such as plant-based proteins, insect meals, and single-cell proteins, that can help reduce dependency on traditional marine resources.

2. Nutritional Requirements of Carnivorous Fish

Carnivorous fish, which include economically important species such as salmon, trout, sea bass, and groupers, have unique nutritional needs due to their reliance on high-protein diets. Their natural diet consists mainly of other animals, making them metabolically adapted to efficiently process protein and fat. Formulating diets for these species requires a deep understanding of their specific nutrient requirements, particularly in relation to protein, lipids, carbohydrates, vitamins, and minerals. Meeting these nutritional needs is essential for optimizing growth, improving feed conversion ratios (FCR), enhancing health, and ensuring overall aquaculture profitability.

2.1. Protein

Protein is the cornerstone of carnivorous fish diets, serving as a primary nutrient for growth, tissue repair, and various metabolic functions. Unlike herbivorous or omnivorous species, which can derive a larger portion of their energy from carbohydrates, carnivorous fish rely heavily on protein for energy. Protein content in feeds for these fish species generally ranges from 30% to 55%, depending on the species, life stage, and farming conditions.

- **Growth and development**: Protein provides the building blocks (amino acids) necessary for muscle development, tissue repair, and growth. High-quality protein sources, such as fishmeal, ensure that fish receive all essential amino acids.
- **Essential amino acids (EAAs)**: Carnivorous fish cannot synthesize certain amino acids, known as essential amino acids, and therefore must obtain them from their diet. Important EAAs for fish include lysine, methionine, arginine, histidine, and tryptophan. Deficiencies in these amino acids can lead to reduced growth, poor feed utilization, and compromised health.
    - For example, lysine is essential for protein synthesis and bone development, while methionine is critical for protein synthesis and acts as a methyl donor in various metabolic pathways.
- **Protein sources**: Traditionally, fishmeal has been the preferred protein source due to its balanced amino acid profile, high digestibility, and palatability. However, due to sustainability and cost concerns, there is a growing focus on alternative protein sources like plant proteins (soy, wheat gluten), insect meals, and single-cell proteins (microalgae,

yeast). These alternatives must be evaluated for amino acid content, digestibility, and potential anti-nutritional factors.

2.2. Lipids

Lipids provide a concentrated and highly efficient source of energy for carnivorous fish, yielding about 9 kcal/g, which is more than double the energy yield of protein or carbohydrates. In addition to being an energy source, lipids are critical for supplying essential fatty acids (EFAs) that fish cannot synthesize on their own.

- **Essential fatty acids (EFAs)**: Omega-3 (n-3) and omega-6 (n-6) fatty acids are crucial for maintaining cell membrane integrity, supporting immune function, and promoting growth and reproduction. EFAs are also involved in inflammatory responses and stress tolerance, which are important for fish health under intensive farming conditions.
    - **Omega-3 fatty acids**, such as eicosapentaenoic acid (EPA) and docosahexaenoic acid (DHA), are abundant in marine fish oils and are vital for brain development, cardiovascular health, and anti-inflammatory effects.
    - **Omega-6 fatty acids**, such as linoleic acid, are found in plant-based oils and are important for promoting growth and immune function. However, the ratio of omega-3 to omega-6 in the diet must be balanced, as excessive omega-6 can promote pro-inflammatory responses.
- **Dietary lipid levels**: Carnivorous fish typically require 10% to 20% lipids in their diet, depending on species, life stage, and water temperature. Excessive dietary fat can lead to fatty liver disease, while insufficient lipid intake may impair growth and reproductive function.
- **Lipid sources**: Fish oil is the traditional lipid source in aquaculture feeds due to its high EPA and DHA content. However, with the increasing cost and sustainability concerns, alternatives such as plant oils (canola, flaxseed, soybean) and algal oils are being explored. These alternatives often lack sufficient levels of EPA and DHA, requiring supplementation or blending with fish oil to achieve the desired fatty acid profile.

2.3. Carbohydrates

Unlike herbivorous or omnivorous fish, carnivorous species have a limited capacity to utilize carbohydrates as an energy source. In their natural diet, which consists mainly of animal prey, carbohydrates are present in very small amounts. However, carbohydrates are included in formulated diets primarily for energy, as well as to serve as binders for feed pellets.

- **Limited carbohydrate utilization**: Carnivorous fish have low levels of digestive enzymes, such as amylase, required to break down complex carbohydrates. As a result, their capacity to digest and metabolize carbohydrates is much lower than that of other fish species. Excessive carbohydrate intake can result in poor growth, reduced feed efficiency, and metabolic disorders, such as impaired glucose tolerance.
- **Dietary inclusion**: Carbohydrate levels in carnivorous fish diets are typically kept low, ranging from 15% to 25% of the total diet. The primary role of carbohydrates in these diets is to provide a cost-effective source of energy, allowing more expensive protein to be spared for growth and tissue maintenance. Sources of carbohydrates in fish feed may include **starches** from grains (corn, wheat, rice) and simple sugars (glucose, dextrins).
- **Carbohydrate quality**: The type of carbohydrate used in the diet can influence digestibility and growth performance. Simple carbohydrates are more easily digested, while complex carbohydrates like fiber may be indigestible and can negatively impact nutrient absorption.

2.4. Vitamins and Minerals

Vitamins and minerals are micronutrients that are essential for the proper physiological functioning of carnivorous fish. They support a range of biological processes, including growth, immune response, skeletal development, and reproduction. Deficiencies or imbalances in vitamins and minerals can lead to poor health, increased disease susceptibility, and impaired growth.

- **Vitamins**: Fat-soluble vitamins (A, D, E, K) and water-soluble vitamins (B-complex, C) are essential for various metabolic processes.
  - **Vitamin A**: Important for vision, growth, and immune function.
  - **Vitamin D**: Essential for calcium and phosphorus metabolism, bone development, and skeletal health.
  - **Vitamin E**: An antioxidant that protects cells from oxidative damage, supports immune function, and enhances reproductive health.
  - **Vitamin K**: Crucial for blood clotting and bone development.
- **Minerals**: Key minerals include calcium, phosphorus, magnesium, and trace elements like zinc, iron, and selenium. These are involved in bone formation, muscle function, enzyme activation, and osmoregulation.
  - **Phosphorus**: Particularly important for carnivorous fish as it supports skeletal development and energy metabolism. It is often provided through inorganic phosphate supplements because plant-based sources of phosphorus are not easily digestible by fish.

3. Feed Ingredients and Their Role in Formulation

The formulation of feed for carnivorous fish species requires selecting ingredients that meet nutritional requirements, enhance growth performance, and minimize costs. Historically, fishmeal and fish oil have been the cornerstones of carnivorous fish diets, providing high-quality protein and essential fatty acids. However, due to sustainability concerns and rising costs, there is a growing shift towards alternative protein and lipid sources, with the use of functional feed additives gaining importance in enhancing nutrient utilization.

3.1. Fishmeal and Fish Oil

Fishmeal and fish oil have long been the preferred ingredients in feed formulations for carnivorous fish, such as salmon, trout, and sea bass, due to their high nutritional value:

- Fishmeal is derived from small, pelagic fish and is rich in essential amino acids, such as lysine, methionine, and histidine, which are critical for muscle growth and metabolic functions in carnivorous species. It is highly digestible and promotes excellent growth and feed conversion ratios (FCRs). However, concerns about overfishing and the rising costs of fishmeal have spurred research into alternative protein sources.
- Fish oil is the primary source of long-chain omega-3 fatty acids, particularly eicosapentaenoic acid (EPA) and docosahexaenoic acid (DHA), which are crucial for immune function, reproduction, and maintaining cellular health. The high demand for fish oil, coupled with limited supply, makes it increasingly expensive and unsustainable.

Challenges with Fishmeal and Fish Oil

- **Sustainability**: The over-reliance on wild fish stocks for fishmeal and fish oil has led to pressure on marine ecosystems, prompting calls for more sustainable alternatives.
- **Cost**: As demand for fishmeal and fish oil increases globally, so do their prices, making feed formulation more expensive and less accessible for small-scale aquaculture operations.

3.2. Alternative Protein Sources

To reduce dependency on fishmeal and lower feed costs, a variety of alternative protein sources have been explored. These sources must be nutritionally balanced and cost-effective while maintaining the performance and health of the fish.

- **Plant-based proteins**: Plant proteins, such as soybean meal, canola meal, pea protein, and wheat gluten, have gained traction as alternatives to fishmeal. Soybean meal is the most commonly used plant protein due to its high protein content and availability. However, plant proteins typically lack certain essential amino acids, such as lysine and methionine, and may contain anti-nutritional factors like phytates, lectins, and protease inhibitors, which can impair nutrient digestibility. Therefore, careful balancing and supplementation with synthetic amino acids or enzyme additives are necessary.
  - **Soybean meal**: Widely used due to its availability and cost-effectiveness, soybean meal provides a moderate protein content (~44%-48%) but requires amino acid supplementation due to deficiencies in methionine.
  - **Canola meal**: A by-product of canola oil extraction, canola meal has a good amino acid profile but lower protein content than fishmeal. It is often used in combination with other ingredients.
  - **Pea protein**: Pea protein isolate has been considered as an alternative to fishmeal, particularly in Europe. However, like other plant proteins, it lacks certain essential amino acids.
- **Insect meal**: Insects, such as black soldier fly larvae (BSFL), mealworms, and crickets, are rich in protein and essential amino acids and are increasingly being explored as sustainable alternatives to fishmeal. Insect meal is not only highly nutritious but also environmentally friendly, as insects can be reared on organic waste. Studies have shown that fish fed insect meal exhibit growth performance and feed efficiency similar to those fed fishmeal-based diets.
- **Animal by-products**: Animal by-products, such as poultry meal, blood meal, and meat and bone meal, provide high-quality protein and are often more affordable than fishmeal. These by-products are used to recycle waste from the meat-processing industry and can help reduce feed costs while maintaining nutritional adequacy. However, the inclusion of animal by-products may raise concerns regarding disease transmission and consumer acceptance in certain regions.

3.3. Alternative Lipid Sources

Like protein, alternative lipid sources are crucial for reducing reliance on fish oil while maintaining the energy and fatty acid needs of carnivorous fish. Lipids provide a concentrated source of energy and supply essential fatty acids that support growth, immunity, and reproduction.

- **Vegetable oils**: Oils derived from soybean, canola, sunflower, and palm are widely used as fish oil substitutes. They are cost-effective and readily available, but they contain higher proportions of omega-6 fatty acids compared to fish oil, which is rich in omega-3 fatty acids. Vegetable oils need to be used in combination with omega-3-rich sources like algal oil or supplemented with synthetic EPA and DHA to ensure a balanced fatty acid profile.
- **Algal oils**: Algal oils are one of the most promising sustainable alternatives to fish oil as they are rich in EPA and DHA, similar to fish oil. However, their current production cost is high, making them less accessible for large-scale use in aquafeeds. Continued research is needed to improve production efficiency and reduce costs.
- **Animal fats**: Poultry fat and tallow from the rendering industry can be used as alternative lipid sources in fish diets. They provide an energy-dense source of lipids and are more cost-effective than fish oil. However, their fatty acid composition may need adjustment through supplementation with omega-3-rich oils.

3.4. Additives

Feed additives play a crucial role in improving the digestibility, palatability, and health benefits of formulated diets. These include:

- **Probiotics and prebiotics**: Probiotics (live beneficial bacteria) and prebiotics (non-digestible carbohydrates that promote the growth of beneficial gut bacteria) help improve gut health and enhance nutrient absorption. Common probiotics used in aquafeeds include Lactobacillus and Bacillus species.
- **Enzymes**: Enzyme additives, such as phytase, protease, and carbohydrase, improve the digestibility of feed ingredients, especially when plant-based proteins are used. Phytase, for instance, breaks down phytate, an anti-nutritional factor in plant meals, thereby increasing phosphorus availability.
- **Antioxidants**: To prevent lipid oxidation and improve the shelf-life of feeds, antioxidants like vitamin E and selenium are commonly added. These antioxidants also play a vital role in supporting the immune system and maintaining fish health.

4. Balancing Performance and Cost in Feed Formulation

Optimizing feed formulation for carnivorous fish involves balancing the need for high performance (growth and health) with cost-effectiveness. This requires selecting ingredients that provide adequate nutrition while minimizing feed costs.

4.1. Formulation Strategies

- **Cost-effective ingredient selection**: Feed formulation strategies often involve a trade-off between the nutritional quality of ingredients and their cost. Combining high-quality ingredients, like fishmeal and fish oil, with more affordable alternatives (e.g., plant proteins, insect meal, or animal by-products) can help achieve a balanced diet at a lower cost.
- **Nutritional profiling**: Conducting detailed nutritional profiling of feed ingredients allows formulators to optimize the inclusion levels of each ingredient while ensuring the diet meets the fish's essential nutrient requirements. This includes analyzing protein and amino acid content, lipid and fatty acid profiles, as well as vitamin and mineral compositions.
- **Utilization of local ingredients**: Utilizing locally available ingredients reduces transportation costs and supports regional agriculture. Ingredients such as locally grown soy or insect meal can be used to replace more expensive, imported ingredients.
- **Phased feeding programs**: Phased feeding programs adjust the nutrient levels in the diet based on the growth stage of the fish. Juveniles require higher protein and lipid content for rapid growth, while adults may benefit from diets lower in protein but higher in energy. Phased feeding allows for more precise nutrient targeting, optimizing growth and minimizing feed wastage.

4.2. Economic Considerations

- **Market fluctuations**: Feed ingredient prices can fluctuate due to market demand, climate conditions, and geopolitical factors. Formulators need to develop flexible strategies that allow for adjustments in ingredient selection based on availability and cost fluctuations.
- **Feed conversion ratio (FCR)**: A key metric for cost-effectiveness in aquaculture is the feed conversion ratio, which measures the amount of feed required to produce a unit of fish weight. Lowering FCR by optimizing diet formulations improves feed efficiency, directly reducing feed costs. Selecting highly digestible ingredients and using additives like enzymes can enhance FCR.

5. Challenges in Feed Formulation for Carnivorous Species

Formulating diets for carnivorous fish species is not without challenges, and aquaculture practitioners must navigate a range of factors that impact both feed efficiency and sustainability. These challenges include ingredient variability, the risk of nutritional imbalances, and compliance with regulatory frameworks, all of which can complicate efforts to create cost-effective yet nutritionally adequate diets.

5.1. Ingredient Variability

The nutritional composition of feed ingredients is subject to variability, which can affect the overall consistency and quality of formulated diets. Factors influencing ingredient variability include:

- **Climate conditions**: Weather patterns, such as droughts or excessive rainfall, can impact the nutrient composition of plant-based ingredients like soybean meal or corn. For instance, soybean meal grown in different regions may vary in protein and amino acid content.
- **Soil conditions**: Soil fertility plays a role in determining the nutritional profile of plant-based ingredients. Poor soil quality may lead to lower nutrient concentrations in crops used in feed formulation.
- **Processing methods**: The method used to process raw ingredients can impact their nutritional value. For example, excessive heat treatment during ingredient processing can reduce the availability of essential amino acids, such as lysine, due to **Maillard reactions**. Similarly, the processing of animal by-products can alter their fat and protein content.

The variability in protein and lipid levels of ingredients can lead to fluctuations in the final feed's nutritional composition, affecting fish growth performance and feed conversion efficiency. Inconsistent ingredient quality may necessitate frequent adjustments in formulation, making it difficult for farmers to maintain a stable diet for their fish.

5.2. Nutritional Imbalances

Balancing essential nutrients while keeping feed costs low is a delicate task that, if not managed properly, can lead to nutritional imbalances. These imbalances can negatively affect fish health and growth performance, leading to economic losses. Specific challenges include:

- **Amino acid imbalances**: Replacing fishmeal with plant-based proteins often results in a diet deficient in certain essential amino acids, such as methionine and lysine. Although these amino acids can be supplemented, improper balancing can lead to poor growth, inefficient feed utilization, or even metabolic disorders in carnivorous fish.
- **Fatty acid imbalances**: Alternative lipid sources, such as vegetable oils, often lack the long-chain omega-3 fatty acids (EPA and DHA) found in fish oil. A deficiency in these critical fatty acids can impair

immune function, reproduction, and growth in carnivorous fish. If fish oil alternatives are not fortified with EPA and DHA, fish performance may suffer.

- **Vitamin and mineral deficiencies**: As fishmeal and fish oil are reduced in diets, attention must be given to ensuring sufficient vitamins and minerals. For instance, plant-based proteins often contain anti-nutritional factors that reduce the bioavailability of key minerals like phosphorus. Deficiencies in these micronutrients can lead to skeletal deformities, poor growth, and reduced disease resistance.

To avoid nutritional imbalances, it is crucial to regularly monitor the nutrient composition of feed ingredients and adjust formulations based on the specific needs of the target fish species. This requires precision in feed formulation and careful consideration of the life stage and health status of the fish.

5.3. Regulatory Constraints

The aquafeed industry operates under a complex regulatory framework that governs the safety, sustainability, and environmental impact of feed ingredients. These regulations can pose challenges for feed formulators, particularly when sourcing novel or alternative ingredients. Some of the key regulatory constraints include:

- **Feed ingredient approval**: In many countries, feed ingredients must be approved by regulatory bodies before they can be used in commercial aquaculture. For example, the European Union (EU) has strict regulations on the use of animal by-products and genetically modified organisms (GMOs) in aquafeeds. This can limit the availability of certain ingredients and increase reliance on traditional, more expensive feed components like fishmeal.
- **Environmental impact assessments**: The environmental sustainability of feed formulation is becoming increasingly important, with regulators requiring aquaculture operations to minimize their environmental footprint. Feed ingredients that contribute to high nitrogen or phosphorus waste output may face restrictions, as excessive nutrient discharge into aquatic environments can lead to eutrophication and ecosystem degradation.
- **Food safety standards**: Regulations related to food safety, such as the presence of contaminants (e.g., heavy metals or pesticides), must be strictly adhered to when selecting feed ingredients. This may limit the use of certain by-products or novel ingredients that do not meet safety standards, despite their potential nutritional or economic benefits.

Navigating these regulatory frameworks requires feed formulators to carefully evaluate ingredient sources and comply with local and international standards, which can add complexity and cost to the feed formulation process.

6. Future Perspectives

The aquaculture industry is under increasing pressure to adopt more sustainable practices, particularly as global demand for seafood rises. Future innovations in feed formulation will focus on improving ingredient sustainability, nutritional efficiency, and cost-effectiveness. Several key areas offer significant potential for advancing feed formulation for carnivorous fish species:

6.1. Innovative Ingredient Development

Research into novel protein and lipid sources will play a pivotal role in reducing the industry's reliance on traditional ingredients like fishmeal and fish oil. Some promising innovations include:

- **Microbial proteins**: Proteins derived from microorganisms, such as bacteria, yeast, or single-cell algae, are emerging as a sustainable and scalable alternative to fishmeal. These **single-cell proteins (SCPs)** offer a high-quality protein source with a balanced amino acid profile. Additionally, SCPs can be produced using waste streams, further enhancing their sustainability.
- **Alternative oils**: The development of **microalgal oils** and **genetically modified crops** rich in omega-3 fatty acids holds great promise for replacing fish oil. Microalgae, which are the primary source of omega-3s in the marine food chain, can be cultured at scale to provide a sustainable and renewable source of EPA and DHA for aquafeeds.
- **Insect proteins**: The use of insect meal in aquafeeds is gaining momentum as a sustainable and nutritionally balanced alternative to fishmeal. The black soldier fly (*Hermetia illucens*) is one of the most studied species, with its larvae providing a high-quality protein and lipid source that can enhance fish growth performance.

6.2. Precision Nutrition

Precision nutrition is an emerging approach that aims to tailor diets based on the specific nutrient requirements of individual fish or fish populations. Advances in omics technologies (genomics, proteomics, and metabolomics) and machine learning can provide insights into the precise nutrient needs of fish at different life stages or under varying environmental conditions.

- **Nutrient modeling**: By combining data from nutrient profiling with growth models, precision nutrition can optimize feed formulations to maximize growth efficiency while minimizing nutrient waste. This approach reduces feed costs and minimizes the environmental impact of aquaculture.

- **Real-time monitoring**: The development of real-time monitoring systems, such as **nanosensors**, can help track nutrient levels in feed and water, allowing for dynamic adjustments to feeding regimes. These systems can improve feed efficiency by delivering the right nutrients at the right time, reducing both overfeeding and underfeeding.

6.3. Environmental Impact Assessments

As sustainability becomes a top priority for the aquaculture industry, future feed formulation practices will increasingly focus on minimizing their environmental impact. This involves conducting environmental impact assessments of feed ingredients and formulating diets that produce minimal waste and pollution.

- **Nutrient recycling**: Incorporating nutrient recycling technologies, such as anaerobic digesters, can help capture and repurpose waste products from aquaculture systems. By converting fish waste into usable feed ingredients, these technologies contribute to circular economy practices.
- **Carbon footprint reduction**: The development of **low-carbon feeds**, which use ingredients with a lower carbon footprint (e.g., plant-based proteins, insect meal, and microbial oils), will be key to reducing the aquaculture industry's overall environmental impact.

7. Conclusion

Feed formulation for carnivorous fish species is a complex process that must balance nutritional adequacy, cost, and sustainability. By exploring innovative ingredient sourcing, embracing precision nutrition, and adhering to environmental and regulatory standards, aquaculture practitioners can enhance the efficiency and sustainability of fish production systems. The future of feed formulation lies in continued research and development, with a focus on creating diets that not only meet the nutritional needs of carnivorous fish but also promote responsible and environmentally sustainable aquaculture practices. Through collaboration between researchers, feed producers, and aquaculture operators, the industry can continue to evolve toward a more sustainable and economically viable future.

References

Barrows, F.T., Gaylord, T.G., & Sealey, W.M. (2010). Alternative Protein Sources for Use in Fishmeal-Free Diets for Carnivorous Fish. Aquaculture Nutrition, 16(2), 71-80.

Barrows, F.T., Stone, D.A.J., & Hardy, R.W. (2007). The Effects of Protein Source and Substitution Level on Growth and Survival of Carnivorous Fish Species. Aquaculture Research, 38(5), 510-517.

Cabral, E.M. et al. (2011). Replacement of Fishmeal by Plant Proteins in the Diet of European Sea Bass (Dicentrarchus labrax). Aquaculture, 315(3), 295-301.

Cheng, Z.J., Hardy, R.W., & Usry, J.L. (2003). Effects of Lysine Supplementation in Plant-Based Diets on the Growth of Juvenile Fish. Aquaculture Nutrition, 9(4), 237-245.

Collins, S.A., et al. (2012). Evaluation of Alternative Lipid Sources in Diets for Carnivorous Fish: Lipid Digestibility and Performance. Aquaculture Nutrition, 18(2), 325-331.

Drew, M.D., Borgeson, T.L., & Thiessen, D.L. (2007). A Review of Processing of Feed Ingredients to Enhance Nutritional Value of Plant Proteins. Animal Feed Science and Technology, 138(2), 170-183.

Francis, G. et al. (2001). The Potential of Plant-Derived Antinutritional Factors in Animal Feed to Promote Growth in Carnivorous Fish. Reviews in Fish Biology and Fisheries, 11(2), 119-173.

Gatlin, D.M. et al. (2007). Expanding the Utilization of Sustainable Plant Products in Aquafeeds: A Review. Aquaculture Research, 38(6), 551-579.

Glencross, B. (2009). Exploring the Nutritional Demand for Essential Amino Acids by Aquaculture Species. Reviews in Aquaculture, 1(2), 112-124.

Glencross, B. (2016). Limitations of Alternative Protein Sources for Aquaculture: The Future of Fishmeal in Aquafeeds. Journal of Aquaculture Research & Development, 47(3), 112-123.

Glencross, B.D. et al. (2017). A Review of the Nutritional and Functional Roles of Fatty Acids in Fish. Aquaculture, 500, 397-407.

Hardy, R.W. (2010). Utilization of Plant Proteins in Fish Diets: Effects of Global Demand and Supplies of Fishmeal. Aquaculture Research, 41(5), 770-776.

Hardy, R.W. (2013). Research and Applications in the Use of Alternative Protein Sources for Aquafeeds. Journal of Applied Aquaculture, 25(2), 156-170.

Hixson, S.M. (2014). Fish Nutrition and Current Issues in Aquaculture: The Balance in Providing Safe, Nutritious Fish and Maintaining Ecosystem Health. Journal of Aquaculture Research & Development, 5(3), 1-10.

Hua, K., & Bureau, D.P. (2009). Development of an Economical and Sustainable Alternative Feed for Carnivorous Fish. Aquaculture, 289(3-4), 193-202.

Kaushik, S.J., & Hemre, G.I. (2008). Plant Ingredients as Sustainable Protein Sources for Aquafeeds. The European Journal of Lipid Science and Technology, 110(9), 780-787.

Krogdahl, Å., et al. (2010). Soybean Meal in Fish Feed: Antinutritional Factors, Recirculating Aquaculture Systems, and Fish Nutrition. Aquaculture Research, 41(3), 348-363.

Li, P., Mai, K., Trushenski, J., & Wu, G. (2009). New Developments in Fish Amino Acid Nutrition: Towards Functional and Environmentally Oriented Aquafeeds. Amino Acids, 37(1), 43-53.

Naylor, R.L., et al. (2009). Feeding Aquaculture in an Era of Finite Resources. Proceedings of the National Academy of Sciences, 106(36), 15103-15110.

NRC (National Research Council) (2011). Nutrient Requirements of Fish and Shrimp. National Academies Press, Washington, DC.

Oliva-Teles, A. (2012). Nutrition and Health of Aquaculture Fish. Journal of Aquaculture Research and Development, S2:003.

Rombenso, A.N. et al. (2013). Effects of Fish Oil Replacement with Soybean Oil in Aquafeeds on Growth, Lipid Metabolism, and Health of Juvenile Fish. Aquaculture Nutrition, 19(3), 310-321.

Sarker, P.K. et al. (2016). Towards Sustainable Aquafeeds: Complete Substitution of Fish Oil with Marine Microalga Schizochytrium in Feed for Juvenile Nile Tilapia (Oreochromis niloticus). PLOS ONE, 11(6), e0156684.

Sealey, W.M., & Gaylord, T.G. (2012). Effects of Fishmeal Replacement with Plant Protein Sources on Growth and Feed Utilization in Carnivorous Fish. Aquaculture, 354-355, 134-141.

Silva, T.S., et al. (2015). The Replacement of Fishmeal by Plant Protein Sources in the Diets of Carnivorous Fish Species. Aquaculture Nutrition, 21(3), 407-423.

Tacon, A.G.J. (1994). Feed Ingredients for Carnivorous Fish Species: Alternatives to Fishmeal and Fish Oil. Aquaculture Nutrition, 1(2), 73-86.

Tocher, D.R. (2015). Omega-3 Long-Chain Polyunsaturated Fatty Acids and Aquaculture in Perspective. Aquaculture, 449, 94-107.

Trushenski, J.T. et al. (2011). Evaluation of Soybean Meal and Other Plant Protein Sources in the Diet of Juvenile Marine Fish. Aquaculture Nutrition, 17(4), 582-595.

Turchini, G.M., Torstensen, B.E., & Ng, W.K. (2009). Fish Oil Replacement in Finfish Nutrition. Reviews in Aquaculture, 1(1), 10-57.

12

Innovative Protein Sources for Aquaculture: Reducing Pressure on Wild Fisheries

1. Introduction

Aquaculture has emerged as one of the fastest-growing sectors of food production worldwide, driven by the increasing demand for fish and seafood as essential sources of protein, healthy fats, and micronutrients. As wild fisheries face overexploitation and environmental degradation, aquaculture is seen as a solution to meet the growing global need for seafood. However, the sustainability of aquaculture itself is under scrutiny due to its reliance on traditional feed ingredients, particularly fishmeal and fish oil derived from wild-caught fish. Fishmeal has long been considered an ideal protein source for aquafeeds because of its excellent amino acid profile, high digestibility, and essential fatty acid content. Yet, the continued use of fishmeal places immense pressure on wild fish stocks, particularly small pelagic species that are harvested specifically for fishmeal production.

The overexploitation of these fish stocks raises significant ecological concerns, threatening marine biodiversity and destabilizing food webs. The reliance on fishmeal is also economically challenging, as fluctuating supply and increasing demand have led to rising costs and price volatility. Moreover, as the aquaculture industry continues to expand, the demand for fishmeal will surpass the sustainable supply, exacerbating these environmental and economic issues. In response to these challenges, researchers and industry stakeholders have been actively seeking sustainable and cost-effective alternatives to fishmeal. These innovative protein sources are not only essential for reducing the environmental impact of aquaculture but also for ensuring the long-term viability of the industry. The development of alternative protein sources must consider several factors, including their nutritional composition, availability, cost-effectiveness, and environmental footprint.

This chapter delves into the array of alternative protein sources that have been proposed and tested for aquaculture feeds, including plant-based proteins, insect-derived meals, microbial proteins (single-cell proteins), and by-products from other industries. Each of these alternatives offers unique benefits and challenges in terms of sustainability, nutritional adequacy, and their potential to reduce dependence on wild-caught fish.

Plant-based proteins, for example, have gained considerable attention due to their wide availability and relatively low cost. However, challenges such as anti-nutritional factors, lower digestibility, and imbalances in essential amino acids must be addressed to optimize their use in aquafeeds. Insect meals, which mimic the natural diets of many fish species, present another promising solution with high protein content and favorable amino acid profiles. Similarly, microbial proteins derived from bacteria, yeast, and algae offer a novel, scalable, and potentially environmentally friendly source of protein for aquaculture. By-products from agriculture and food industries, such as poultry by-product meal and plant processing residues, also represent valuable protein sources that can reduce waste and improve the sustainability of aquafeeds. Utilizing these by-products not only lowers the cost of feed production but also contributes to the circular economy by repurposing waste materials.

The growing portfolio of alternative protein sources underscores the need for a multifaceted approach to feed formulation that balances sustainability, fish health, and economic feasibility. Incorporating these innovative ingredients into aquaculture feeds can significantly reduce the industry's reliance on fishmeal and help mitigate the ecological impacts of overfishing. Furthermore, the successful integration of alternative proteins into aquafeeds requires continuous research, innovation, and collaboration between researchers, feed manufacturers, and the aquaculture industry.

This chapter provides a comprehensive overview of these alternative protein sources, highlighting their potential to support the sustainable growth of aquaculture. It examines their nutritional value, digestibility, economic viability, and environmental impact, offering insights into how these alternatives can contribute to a more sustainable and resilient aquaculture industry. Through the adoption of innovative protein sources, aquaculture has the opportunity to alleviate pressure on wild fisheries, reduce its environmental footprint, and pave the way for a more sustainable future in global seafood production.

2. The Fishmeal Dilemma

Fishmeal, a high-quality protein source, has long been a cornerstone of aquafeeds, particularly for carnivorous species such as salmon, trout, and seabass. Derived primarily from small pelagic fish species like anchovy, sardine, and menhaden, fishmeal offers an ideal balance of essential amino acids, fatty acids, and vital micronutrients. Its unique nutritional profile makes it indispensable for supporting the growth, health, and reproduction of farmed fish. However, despite these benefits, the reliance on fishmeal presents significant challenges across environmental, economic, and sustainability dimensions. These challenges are driving the search for more sustainable alternatives in aquaculture feed formulation.

2.1. Environmental Impact

The environmental consequences of fishmeal production are one of the most critical concerns. The small pelagic fish species used to produce fishmeal are integral components of marine ecosystems, playing key roles in food webs by serving as prey for larger fish, marine mammals, and seabirds. Overharvesting of these species to meet the growing demand for fishmeal can lead to population declines, potentially destabilizing marine ecosystems.

Several regions where fishmeal is produced, such as Peru and Chile (which supply much of the global market through their anchovy fisheries), have experienced fluctuating fish populations due to overfishing, climate variability (e.g., El Niño events), and ocean temperature changes. Overfishing can lead to the collapse of fish stocks, affecting not only biodiversity but also the livelihoods of local fishing communities that depend on these resources. In addition, the high concentration of fishmeal production in specific regions raises concerns about localized depletion of marine resources, leading to cascading effects throughout the food chain. Furthermore, fishing for small pelagic species contributes to the depletion of fish populations that are already under pressure from climate change and habitat degradation. This exacerbates the strain on marine ecosystems, threatening the long-term sustainability of fisheries and reducing the resilience of ocean ecosystems.

2.2. Economic Concerns

The economic aspect of fishmeal reliance is equally problematic. Fishmeal prices are highly volatile due to a range of factors, including fluctuating fish stocks, regulatory restrictions, and changing environmental conditions. Events such as overfishing or unfavorable climate phenomena like El Niño can significantly reduce the supply of small pelagic species, driving up prices and making fishmeal less affordable for fish farmers. For aquaculture producers, price volatility adds uncertainty to feed production costs, which typically represent more than 50% of the total cost of aquaculture operations. Fishmeal price increases can squeeze profit margins and make aquaculture less competitive compared to other animal protein production systems. As global demand for seafood continues to rise, the pressure on aquafeed producers to maintain competitive prices while ensuring the quality and sustainability of their products becomes more intense.

Another economic consideration is the competition for fishmeal with other industries, such as livestock and poultry production. Fishmeal is also used as a feed ingredient in these sectors, further contributing to demand pressures and price fluctuations. This inter-industry competition exacerbates the supply-demand imbalance, making fishmeal less reliable as a long-term protein source for aquaculture.

2.3. Sustainability Challenges

The growing aquaculture industry, which surpassed wild capture fisheries in global production in 2014, cannot rely solely on fishmeal to sustain its expansion. As aquaculture continues to play a pivotal role in meeting the protein needs of a growing global population, it becomes clear that fishmeal is not a scalable solution for the industry's future. The limited availability of small pelagic fish stocks, coupled with increasing demand from both aquaculture and terrestrial livestock industries, creates a "fishmeal bottleneck" that threatens the sustainability of aquaculture growth. The potential for ecological collapse and economic instability underscores the need to reduce reliance on fishmeal and seek out alternative protein sources. Additionally, the aquaculture industry's increasing focus on environmental sustainability is driving efforts to reduce the carbon footprint and ecological impact of fish farming operations. As fishmeal production is associated with significant environmental costs, including the carbon emissions related to fishing and processing, the industry faces growing pressure from regulatory bodies, consumers, and sustainability certifications (e.g., ASC, MSC) to transition toward more sustainable feed ingredients.

2.4 Social and Ethical Considerations

In addition to the environmental and economic challenges, there are social and ethical considerations associated with fishmeal production. The small pelagic fish species used for fishmeal are often important sources of food and income for coastal communities, particularly in developing countries. Diverting these fish from direct human consumption to industrial aquafeed production can raise concerns about food security and equitable resource distribution. As global demand for seafood grows, there is an increasing ethical responsibility to ensure that marine resources are managed in a way that supports both aquaculture production and the food security needs of vulnerable populations. Reducing the reliance on fishmeal through the adoption of alternative protein sources can help alleviate some of these social pressures and promote a more equitable use of marine resources.

2.5. The Need for Alternative Protein Sources

Given the environmental, economic, and social challenges associated with fishmeal production, the aquaculture industry is actively seeking alternative protein sources that can either replace or significantly reduce the reliance on fishmeal. These alternatives must not only be nutritionally adequate to support fish growth and health but also economically viable, scalable, and environmentally sustainable.

Plant-based proteins, insect meals, microbial proteins (single-cell proteins), and by-products from other industries are among the most promising alternatives being explored. These protein sources offer several advantages, including lower environmental impacts, reduced pressure on wild fisheries, and potential cost savings. However, their incorporation into aquafeeds presents its own set of challenges, such as anti-nutritional factors in plant proteins, digestibility concerns, and consumer acceptance issues. The development and optimization of alternative protein sources represent a crucial step toward a more sustainable future for aquaculture. As research progresses, it is expected that the industry will continue to diversify its feed ingredients, moving away from fishmeal dependency and towards a more sustainable and resilient feed system.

3. Alternative Protein Sources

As global aquaculture expands and the limitations of fishmeal become increasingly apparent, alternative protein sources are being actively explored to reduce the industry's dependency on wild fisheries. These alternative proteins aim to provide a sustainable, cost-effective solution without compromising the nutritional quality of aquafeeds. This section discusses the most promising alternatives, including plant-based proteins, insect meals, microbial proteins, and by-products from other industries.

3.1. Plant-Based Proteins

Plant proteins have become a leading option for replacing fishmeal due to their availability, cost-effectiveness, and environmental sustainability. Key plant-based alternatives include soybean meal, pea protein, canola meal, and corn gluten meal. However, plant proteins often require supplementation or processing to address nutrient imbalances and anti-nutritional factors (ANFs).

3.1.1. Soybean Meal

Soybean meal is one of the most widely used plant proteins in aquafeeds due to its high protein content (typically 44-48%) and broad availability. Soybeans are rich in essential amino acids such as lysine but are deficient in methionine, which must be supplemented in carnivorous fish diets. Despite its extensive use, soybean meal contains several ANFs such as protease inhibitors, lectins, and phytic acid, which can impair nutrient absorption, reduce feed efficiency, and lead to gut inflammation in some species. Innovative processing techniques such as fermentation, extrusion, and enzymatic treatments are being used to reduce these ANFs and improve the digestibility of soybean meal. Fermentation, for example, has been shown to enhance protein digestibility and bioavailability by breaking down anti-nutritional compounds, making fermented soybean meal a more suitable alternative for carnivorous species like salmon and trout.

3.1.2. Pea Protein and Other Legumes

Pea p.rotein and other legume-based ingredients, including lupin and faba bean, are gaining attention as alternatives to soybean meal. Pea protein offers a high lysine content and lower levels of ANFs compared to soybean, which makes it more digestible for many fish species. However, like soybean, pea protein is deficient in methionine, which limits its use as a complete replacement for fishmeal. The use of pea protein in aquafeeds is still relatively limited compared to soybean, primarily due to its lower global availability and higher cost. Nevertheless, its digestibility and potential for crop diversification make pea protein a promising alternative for inclusion in aquafeeds, particularly in combination with other plant proteins.

3.1.3. Canola Meal

Canola meal, a by-product of oil extraction from canola seeds, is another plant-based protein source gaining prominence in aquaculture. Canola meal has a balanced amino acid profile, with higher levels of sulfur-containing amino acids like methionine and cysteine than soybean meal. It also contains significant levels of crude protein (36-40%), making it a valuable ingredient for aquafeeds. However, canola meal contains ANFs such as glucosinolates, which can interfere with fish metabolism and thyroid function if present in high concentrations. Advances in plant breeding and processing have reduced glucosinolate levels in modern canola cultivars, making canola meal a more viable option for inclusion in aquafeeds for species like tilapia and trout.

3.2. Insect-Based Protein

Insects represent a novel and highly sustainable source of protein for aquaculture. The rapid growth cycle, ability to utilize organic waste, and high nutritional value of insect meals make them an attractive alternative to fishmeal. Among the most researched insect species are black soldier fly larvae (BSFL), mealworms, and crickets.

3.2.1. Black Soldier Fly Larvae (BSFL)

BSFL are particularly promising due to their high protein content (40-45%) and favorable amino acid profile, which is comparable to that of fishmeal. BSFL also contain essential lipids, including lauric acid, which may offer antimicrobial benefits to farmed fish. BSFL can be reared on a wide range of organic wastes, reducing the environmental footprint of feed production and contributing to circular economy models. Research has shown that BSFL meal can replace up to 50% of fishmeal in the diets of species such as salmon, trout, and tilapia without negatively impacting growth performance or feed conversion efficiency. Additionally, BSFL meal contains bioactive compounds such as antimicrobial peptides and chitin, which may enhance the immune system of fish, making it a functional feed ingredient.

3.2.2. Mealworms

Mealworms (Tenebrio molitor) are another insect-based protein source with significant potential for aquaculture. Mealworm meal is rich in protein (50-55%) and has a well-balanced amino acid profile. Studies on species such as rainbow trout and European sea bass have demonstrated that mealworm meal can replace a portion of fishmeal in aquafeeds without compromising growth rates or feed efficiency. The primary challenge with mealworms is scaling up production to meet the demand for aquafeeds. While mealworm farming is relatively resource-efficient, it is not yet as widely adopted as BSFL production. Advances in insect farming technologies and economies of scale are expected to drive down costs and increase the availability of mealworm meal for aquaculture.

3.3. Microbial Proteins

Microbial proteins, also known as single-cell proteins, are derived from microorganisms such as bacteria, yeast, fungi, and algae. These proteins can be produced using a variety of substrates, including industrial by-products, waste materials, and even carbon dioxide, making them highly sustainable and scalable options for aquaculture.

3.3.1. Yeast

Yeast-based proteins, particularly from *Saccharomyces cerevisiae*, have shown promise as fishmeal replacements in aquafeeds. Yeast proteins are rich in essential amino acids and contain bioactive compounds like β-glucans, which can stimulate the immune system of fish. Yeast protein concentrates have been successfully used to replace up to 50% of fishmeal in the diets of species such as rainbow trout and Atlantic salmon, with no adverse effects on growth or feed conversion.

3.3.2. Microalgae

Microalgae such as *Spirulina* and *Chlorella* are rich in protein, omega-3 fatty acids, vitamins, and pigments. They offer significant potential as alternative feed ingredients, especially for species with high nutritional requirements. However, the production of microalgae is currently limited by high costs, primarily due to the energy-intensive cultivation and harvesting processes. Innovations in photobioreactor design, genetic engineering, and large-scale algal farming could help reduce production costs and increase the feasibility of microalgae as a fishmeal alternative.

3.4. By-Products from Other Industries

By-products from the poultry, meat, and plant processing industries are increasingly being utilized as cost-effective and sustainable protein sources in aquafeeds. These by-products provide a way to recycle nutrients and reduce waste, contributing to circular economy practices.

3.4.1. Poultry By-Product Meal

Poultry by-product meal, produced from the rendering of poultry carcasses, is a high-protein feed ingredient that has been successfully used in aquafeeds for species such as salmon, sea bass, and tilapia. Poultry by-product meal offers a good balance of essential amino acids and is more cost-effective than fishmeal. However, concerns over digestibility, palatability, and the potential transmission of pathogens limit its use in some markets.

3.4.2. Feather Meal

Feather meal, produced from hydrolyzed poultry feathers, is another by-product with potential as a protein source for aquaculture. While it has a high crude protein content (80-85%), its amino acid profile is imbalanced, with low levels of lysine and methionine. Enzymatic treatments have been shown to improve the digestibility of feather meal, making it a viable option for some fish species when used in combination with other protein sources.

3.4.3. Plant Processing By-Products

By-products from plant processing, such as rice bran, wheat middlings, and potato protein, offer additional opportunities to incorporate plant-based proteins into aquafeeds. Although these by-products are lower in protein compared to fishmeal, they can be used to supplement other protein sources, particularly in the diets of omnivorous and herbivorous species like tilapia and carp. Their cost-effectiveness and availability make them attractive options for reducing feed costs.

4. Balancing Performance and Sustainability

The transition from traditional fishmeal-based aquafeeds to alternatives poses significant challenges, particularly in balancing fish growth performance, feed conversion efficiency (FCE), and sustainability. While alternative protein sources such as plant proteins, insect meals, and microbial proteins show promise in replacing fishmeal, each comes with limitations that must be addressed to maintain fish health and growth.

4.1. Nutritional Gaps in Alternative Proteins

One of the primary challenges of using alternative protein sources is their incomplete amino acid profiles. Fishmeal is rich in essential amino acids, especially methionine and lysine, which are crucial for the growth and health of carnivorous fish species. In contrast, many plant-based proteins, such as soybean and pea protein, are deficient in one or more of these amino acids. To achieve the same nutritional balance provided by fishmeal, careful formulation is required. This often involves the supplementation of synthetic amino acids or the blending of different protein sources to meet the specific nutritional needs of the fish. For instance, soybean meal is rich in lysine but lacks sufficient methionine, while canola meal contains methionine but is lower in lysine. By combining these two plant-based proteins, aquafeed producers can create a more balanced amino acid profile. However, such formulations must be done with precision to avoid nutrient imbalances that could negatively affect fish growth and FCE.

4.2. Digestibility and Palatability Concerns

The digestibility and palatability of alternative protein sources are also critical factors in aquafeed formulation. Fishmeal has high digestibility across a wide range of species, whereas plant proteins like soybean meal contain anti-nutritional factors (ANFs) such as protease inhibitors, phytates, and lectins, which can reduce nutrient absorption and lead to poor FCE. Advances in feed processing, such as fermentation and enzymatic treatments, have been successful in mitigating the effects of these ANFs, improving the digestibility of plant proteins.

Insect meals, particularly from black soldier fly larvae (BSFL), offer high digestibility and a favorable amino acid profile for many fish species. However, species-specific differences in the ability to digest insect proteins mean that insect meals cannot be universally applied across all aquaculture species without modification. For example, while BSFL meal is well tolerated by species like tilapia and trout, carnivorous fish such as salmon may require additional supplementation or processing to improve digestibility.

Palatability is another critical consideration. Some alternative protein sources, such as plant proteins and by-products from the poultry industry, may be less palatable to fish, leading to reduced feed intake and slower growth. This can be mitigated by incorporating attractants or flavor enhancers into the feed, but these solutions may increase production costs.

4.3. Blended Protein Sources for Optimal Performance

Given the limitations of individual protein sources, a blended approach is often necessary to optimize both performance and sustainability in aquafeeds. By combining plant-based proteins with insect meals, microbial proteins, and by-products from other industries, aquafeed producers can create more balanced, nutritionally complete diets that reduce the reliance on fishmeal. This multi-protein strategy not only helps address the amino acid deficiencies inherent in many alternative proteins but also improves feed digestibility and palatability. For example, a diet for carnivorous fish could include a blend of soybean meal (for lysine), canola meal (for methionine), BSFL meal (for overall protein quality and lipids), and yeast protein (for immune-boosting β-glucans). This approach, combined with targeted supplementation of essential amino acids, can ensure that fish receive the nutrients they need for optimal growth and health, while minimizing the environmental and economic impacts of fishmeal production.

4.4. Sustainability Considerations

Sustainability is at the heart of the shift towards alternative protein sources. The overreliance on fishmeal not only depletes wild fish stocks but also contributes to habitat destruction and biodiversity loss. By incorporating more sustainable protein sources, aquaculture can reduce its ecological footprint while supporting global food security. Insect-based proteins are particularly appealing from a sustainability perspective due to their ability to be farmed on organic waste, significantly reducing resource inputs. Similarly, microbial proteins, produced using industrial by-products or even carbon dioxide, offer a promising solution for reducing the carbon footprint of aquafeed production.

However, it is essential to balance sustainability goals with economic viability. While alternative proteins may reduce the environmental impact of aquafeed production, they can be more expensive to produce than traditional fishmeal, particularly if advanced processing methods are required. This makes it essential for the aquaculture industry to continue investing in research and development to lower production costs and improve the scalability of these protein sources.

5. Future Perspectives and Innovations

The future of aquafeed development lies in continuous innovation, with biotechnology playing a central role in the creation of novel, sustainable protein sources. Precision fermentation, in particular, holds significant potential. This process involves using genetically engineered microbes to produce high-quality, tailored proteins that meet the specific nutritional needs of fish. These proteins can be produced using low-cost substrates, offering a scalable and environmentally friendly alternative to fishmeal.

Advances in processing technologies, such as enzymatic treatments and fermentation, also promise to enhance the digestibility and nutritional value of alternative proteins. For example, fermenting plant proteins can break down anti-nutritional factors and improve amino acid availability, making them more suitable for carnivorous fish species. Similarly, enzymatic treatments can improve the digestibility of insect meals and by-products like feather meal, making them more viable for inclusion in commercial aquafeeds.

Additionally, research into new feed ingredients, such as algae-based proteins and synthetic amino acids, will further diversify the protein sources available for aquafeeds. Algae, in particular, offers a rich source of protein, omega-3 fatty acids, and other essential nutrients, but its high production costs currently limit its use. Innovations in cultivation techniques, such as photobioreactors and genetic engineering, could lower these costs and make algae a more accessible feed ingredient. Collaboration between industry, academia, and policymakers will be critical in driving these innovations forward. Governments and regulatory bodies can support sustainable feed practices by providing incentives for the adoption of alternative proteins and funding research into novel feed ingredients. At the same time, industry stakeholders must continue to invest in research and development to scale up the production of alternative proteins and integrate them into commercial aquafeeds.

6. Conclusion

The future of aquaculture depends on reducing its reliance on fishmeal and adopting more sustainable protein sources. Plant proteins, insect meals, microbial proteins, and industry by-products each offer viable alternatives, but no single solution can meet the nutritional needs of all farmed fish species. A multifaceted approach to feed formulation, combining multiple protein sources and supplementing with essential amino acids, offers the best path forward for maintaining fish growth performance and feed efficiency while promoting sustainability. Innovative protein sources and feed technologies will play a vital role in the sustainable growth of the aquaculture industry, ensuring that farmed fish can continue to meet global demand for high-quality seafood without compromising the health of marine ecosystems. By embracing these innovations, aquaculture can contribute to global food security and environmental sustainability for generations to come.

References

Barroso, F.G., de Haro, C., Sánchez-Muros, M.J., Venegas, E., Martínez-Sánchez, A., & Pérez-Bañón, C. (2014). The potential of various insect species for use as food for fish. Aquaculture, 422-423, 193-201.

Basto, A., Matos, E., Valente, L.M.P., & Neto, A.I. (2020). Growth performance, tissue composition, and liver metabolic enzymes of juvenile sea bass fed diets with increasing levels of seaweed. Journal of Applied Phycology, 32(5), 3171-3181.

Belghit, I., Lock, E.J., Fountoulaki, E., & Johansen, J. (2020). Insect-based ingredients in aquafeeds: Insights into dietary use and future prospects. Reviews in Aquaculture, 12(4), 1836-1850.

Chia, S.Y., Tanga, C.M., Osuga, I.M., Cheseto, X., Ekesi, S., & Van Loon, J.J.A. (2019). Nutritional composition of black soldier fly larvae feeding on agro-industrial by-products. Entomologia Experimentalis et Applicata, 167(3), 244-255.

Daniel, N., Kumar, V., Kumar, S., Verma, A.K., & Roy, D. (2018). Insect meal as an emerging source of protein for fish and poultry: A review. Indian Journal of Animal Research, 52(3), 339-347.

Davies, S.J., Gouveia, A., Laporte, J., Woodgate, S.L., & Dick, J.R. (2011). Nutrient digestibility profile of processed animal proteins supporting the application of products within aquafeeds. Aquaculture Research, 42(6), 797-810.

Gasco, L., Finke, M., & van Huis, A. (2018). Can diets containing insects promote animal health? Journal of Insects as Food and Feed, 4(1), 1-4.

Gatlin, D.M., Barrows, F.T., Brown, P., Dabrowski, K., Gaylord, T.G., Hardy, R.W., Herman, E., Hu, G., Krogdahl, A., Nelson, R., Overturf, K., Rust, M., Sealy, W., Skonberg, D., Souza, E.J., Stone, D., Wilson, R., & Wurtele, E. (2007). Expanding the utilization of sustainable plant products in aquafeeds: A review. Aquaculture Research, 38(6), 551-579.

Glover, M.J., & Mortimer, R. (2021). The role of fermentation in enhancing the nutritional properties of plant-based aquafeed ingredients: A review. Animal Feed Science and Technology, 275, 114891.

Hardy, R.W. (2010). Utilization of plant proteins in fish diets: Effects of global demand and supplies of fishmeal. Aquaculture Research, 41(5), 770-776.

Henry, M., Gasco, L., Piccolo, G., & Fountoulaki, E. (2015). Review on the use of insects in the diet of farmed fish: Past and future. Animal Feed Science and Technology, 203, 1-22.

Hua, K., & Bureau, D.P. (2012). Modelling digestible phosphorus content of salmonid fish feeds. Aquaculture, 350-353, 63-73.

Jones, S.W., Karpol, A., Friedman, S., Maru, B.T., & Tracy, B.P. (2020). Recent advances in single cell protein use as a feed ingredient in aquaculture. Current Opinion in Biotechnology, 61, 189-197.

Kaushik, S.J., & Hemre, G.I. (2008). Plant proteins as alternatives for fishmeal in diets for farmed fish. Journal of Aquaculture Nutrition, 14(3), 99-104.

Kumar, V., Makkar, H.P.S., & Becker, K. (2011). Detoxified Jatropha curcas kernel meal as a dietary protein source: Growth performance, nutrient utilization and digestive enzymes in common carp (Cyprinus carpio L.) fingerlings. Aquaculture Nutrition, 17(3), 313-326.

Li, Y., Bruni, L., Jaramillo-Torres, A., Gai, F., Gasco, L., & Secombes, C.J. (2019). Response of the gut immune system to dietary black soldier fly (Hermetia illucens) larvae meal in rainbow trout. Fish & Shellfish Immunology, 86, 1106-1113.

Makkar, H.P.S., Tran, G., Heuzé, V., & Ankers, P. (2014). State-of-the-art on use of insects as animal feed. Animal Feed Science and Technology, 197, 1-33.

Naylor, R.L., Hardy, R.W., Bureau, D.P., Chiu, A., Elliott, M., Farrell, A.P., Forster, I., Gatlin, D.M., Goldburg, R.J., Hua, K., & Nichols, P.D. (2009). Feeding aquaculture in an era of finite resources. Proceedings of the National Academy of Sciences, 106(36), 15103-15110.

Olsen, R.L., & Hasan, M.R. (2012). A limited supply of fishmeal: Impact on future increases in global aquaculture production. Trends in Food Science & Technology, 27(2), 120-128.

Rahimnejad, S., Lee, S.M., Choi, J., Park, J., & Kim, K.W. (2019). Effects of partial substitution of fishmeal with fermented soybean meal in diets for Japanese seabass. Aquaculture Nutrition, 25(3), 689-699.

Renna, M., Schiavone, A., Gai, F., Dabbou, S., Lussiana, C., Malfatto, V., Prearo, M., Capucchio, M.T., Biasato, I., Biasibetti, E., De Marco, M., Brugiapaglia, A., Zoccarato, I., Gasco, L. (2017). Evaluation of the suitability of a partially defatted black soldier fly (Hermetia illucens L.) larvae meal as ingredient for rainbow trout diets. Journal of Animal Science and Biotechnology, 8(1), 57.

Sealey, W.M., Gaylord, T.G., Barrows, F.T., Tomberlin, J.K., McGuire, M.A., Ross, C., & St-Hilaire, S. (2011). Sensory analysis of rainbow trout, Oncorhynchus mykiss, fed enriched black soldier fly pre-pupae, Hermetia illucens. Journal of the World Aquaculture Society, 42(1), 34-45.

Sørensen, M., Gong, Y., Bjarnason, F., Vasanth, G., Dahle, D., & Huntley, M. (2017). Microalgae as a source of protein and bioactive compounds in feed for salmonid fish. Journal of Applied Phycology, 29(5), 1967-1973.

Tacon, A.G.J., & Metian, M. (2008). Global overview on the use of fishmeal and fish oil in industrially compounded aquafeeds: Trends and future prospects. Aquaculture, 285 (1-4), 146-158.

Turchini, G.M., Torstensen, B.E., & Ng, W.K. (2009). Fish oil replacement in finfish nutrition. Reviews in Aquaculture, 1(1), 10-57.

13

Future Directions in Fisheries Nutrition Challenges and Opportunities

Introduction

Over the past few decades, fisheries nutrition has undergone a remarkable transformation, driven by the rapid expansion of global aquaculture and the increasing need for sustainable seafood production. As wild fish stocks have become overexploited and the demand for protein-rich aquatic food sources continues to rise, aquaculture has emerged as a vital solution for meeting the growing global need for fish and seafood. By 2014, aquaculture production had surpassed wild-caught fisheries in providing fish for human consumption, marking a significant shift in how we obtain seafood. This growth, however, brings with it numerous challenges, particularly regarding the nutrition of farmed fish.

Fisheries nutrition, as a specialized field, plays a pivotal role in supporting aquaculture's ability to provide healthy, high-quality fish for human consumption while maintaining environmental balance. As aquaculture intensifies, the demand for high-quality, cost-effective feed ingredients has surged. Traditional reliance on marine-derived ingredients like fishmeal and fish oil is becoming increasingly unsustainable, as these resources are finite and face depletion due to overfishing. This has resulted in rising costs for these inputs, placing economic pressures on aquaculture producers. Additionally, marine-derived ingredients contribute to ecological imbalances, prompting the need for alternative solutions in feed formulation.

At the same time, the environmental impacts of aquafeeds have garnered increased scrutiny. The production of feed ingredients such as soybeans, corn, and other crops can contribute to deforestation, water pollution, and greenhouse gas emissions, undermining efforts to achieve environmental sustainability in aquaculture. Nutrient waste from uneaten feed and fish excreta can also lead to eutrophication in aquatic ecosystems, damaging water quality and biodiversity. Therefore, fisheries nutrition must address not only the nutritional requirements of fish but also the broader ecological footprint of aquafeed production and usage.

In addition to environmental concerns, fisheries nutrition must balance nutritional efficiency with economic viability and animal welfare. Fish, particularly carnivorous species such as salmon and trout, require high-quality protein and lipid sources to maintain optimal growth, feed conversion efficiency (FCR), and health. However, as alternative protein sources are incorporated into aquafeeds to reduce reliance on fishmeal, these feeds may lack essential amino acids, micronutrients, or digestibility properties found in traditional marine ingredients. Poorly formulated diets can compromise fish health, resulting in slower growth, weakened immune systems, and increased susceptibility to disease outbreaks—outcomes that are detrimental to both farmed fish populations and the aquaculture industry's productivity.

The future of fisheries nutrition lies in addressing these interrelated challenges: the rising cost of feed ingredients, the environmental impacts of feed production, and the need for nutritionally balanced, efficient diets that support both fish welfare and the industry's profitability. The development of alternative protein sources—such as plant-based proteins, insect meals, microbial proteins, and by-products from other industries—offers a promising way forward. These alternatives, coupled with advances in biotechnology, feed processing, and precision nutrition, hold the potential to revolutionize the way we feed farmed fish while minimizing environmental impact.

However, the transition to more sustainable and innovative feed solutions requires coordinated efforts across multiple sectors. Industry, academia, and policymakers must collaborate to advance research, develop new feed ingredients, and create regulatory frameworks that support sustainable feed practices. Investments in scientific research to better understand the nutritional requirements of different fish species, combined with a commitment to reducing the ecological footprint of feed production, will be essential for the long-term success of fisheries nutrition.

This chapter delves into the major challenges and opportunities facing fisheries nutrition as the aquaculture sector continues to evolve. By exploring the latest innovations in feed ingredients, functional feeds, and sustainability practices, it outlines how the future of fisheries nutrition can contribute to achieving global food security, ensuring environmental sustainability, and improving the welfare of farmed fish. The future of aquaculture hinges on the ability to overcome these challenges, and the role of nutrition is central to determining the industry's success in the decades to come.

1. Challenges in Fisheries Nutrition

The field of fisheries nutrition faces numerous challenges as aquaculture continues to expand and diversify to meet global seafood demand. The industry must balance the nutritional needs of farmed fish, environmental sustainability, and economic viability. These challenges are intricately linked, as the reliance on traditional feed ingredients like fishmeal and fish oil is becoming unsustainable due to ecological, economic, and supply constraints. Below are key challenges that highlight the complexity of fisheries nutrition:

1.1. Dependence on Marine-Derived Ingredients

Fishmeal and fish oil have been the cornerstone of aquaculture feeds, particularly for carnivorous species such as salmon, trout, and sea bass. These marine-derived ingredients are highly valued due to their excellent nutritional profile, rich in essential amino acids, long-chain omega-3 fatty acids (EPA and DHA), and micronutrients vital for fish growth, immune function, and reproduction. However, the heavy dependence on fishmeal and fish oil poses several challenges:

- **Overfishing and Ecosystem Disruption**: The production of fishmeal and fish oil primarily depends on small pelagic fish species like anchovies, sardines, and menhaden. These species play crucial roles in marine ecosystems, serving as a food source for larger predators such as seabirds, marine mammals, and commercially valuable fish species. The increasing demand for fishmeal and fish oil has led to overfishing of these species, threatening marine biodiversity and disrupting food webs.
- **Supply Volatility**: The supply of marine-derived ingredients is subject to natural fluctuations driven by environmental conditions such as El Niño events, climate change, and changes in fish population dynamics. Regulatory measures, such as fishing quotas and marine conservation policies, also influence the availability of these resources. These fluctuations can cause price volatility, creating uncertainty for aquaculture producers who rely on stable feed costs to maintain profitability.
- **Rising Costs**: The depletion of wild fish stocks, coupled with growing demand from the expanding aquaculture industry, has driven up the prices of fishmeal and fish oil. This trend places financial pressure on fish farmers, especially in developing countries where access to cost-effective feeds is critical for sustaining production. The rising cost of these ingredients necessitates the exploration of alternative protein and lipid sources that are both nutritionally adequate and economically viable.

1.2 Nutritional Imbalance in Alternative Protein Sources

To address the challenges posed by fishmeal and fish oil reliance, numerous alternative protein sources have been explored, including plant-based proteins, insect meals, and microbial proteins. While these alternatives offer sustainability benefits, they present significant nutritional challenges:

- **Incomplete Amino Acid Profiles**: Many plant-based proteins, such as soybean meal, pea protein, and corn gluten, lack one or more essential amino acids that are critical for fish health and growth. For example, soybean meal is deficient in methionine, while pea protein lacks both methionine and cysteine. This imbalance requires the addition of synthetic amino acids or complementary ingredients, increasing formulation complexity and cost.
- **Anti-Nutritional Factors (ANFs)**: Plant-based proteins, especially soybean meal and rapeseed meal, contain ANFs such as protease inhibitors, phytic acid, and lectins, which can impair nutrient absorption, reduce feed efficiency, and negatively affect fish health. Although processing methods like fermentation and enzymatic treatments can reduce the impact of ANFs, these methods add to production costs and may not be feasible on a large scale.
- **Digestibility and Palatability Issues**: The digestibility of alternative protein sources varies across fish species, particularly for carnivorous fish that have evolved to consume high-protein, animal-based diets. Plant proteins often have lower digestibility than fishmeal, leading to reduced feed conversion ratios (FCR) and growth rates. Insects, while promising, can have variable digestibility depending on processing methods. Furthermore, palatability is another concern, as some alternative proteins are less appealing to fish, leading to lower feed intake and growth performance.

1.3. Sustainability and Environmental Impact

Sustainability is a key driver of innovation in fisheries nutrition, yet aquafeed production itself can have significant environmental impacts. These impacts must be carefully managed to ensure the long-term viability of aquaculture as a sustainable food production system:

- **Environmental Footprint of Feed Ingredients**: The cultivation of plant-based protein sources like soybeans and corn requires large amounts of land, water, and agrochemicals. In regions such as the Amazon rainforest, soybean production has contributed to deforestation, biodiversity loss, and increased greenhouse gas emissions. Insect meals and microbial proteins, while more sustainable, face challenges in scaling up production to meet global demand, and the environmental impacts of their production processes are still being assessed.

- **Nutrient Pollution and Eutrophication**: One of the significant environmental challenges associated with aquaculture is nutrient pollution from uneaten feed and fish waste. High levels of nitrogen and phosphorus in effluents can lead to eutrophication of surrounding water bodies, causing harmful algal blooms and oxygen depletion, which can devastate aquatic ecosystems. Reducing nutrient losses from feeds, either through improved feed formulations or feeding practices, is essential to minimize the environmental impact of aquaculture.
- **Balancing Sustainability with Fish Health and Growth**: The need for sustainable feed ingredients must be balanced with maintaining fish health and productivity. Fish require nutritionally balanced diets to grow efficiently and resist disease. As alternative protein sources are integrated into aquafeeds, careful attention must be given to ensuring that these diets meet the specific nutritional needs of each fish species. Poorly formulated diets can compromise fish welfare, increase disease susceptibility, and reduce farm productivity, negating the environmental benefits of alternative feed ingredients.

1.4. Economic Viability and Access to Feed Ingredients

The economic challenge in fisheries nutrition is twofold: providing affordable feed ingredients for farmers and ensuring that feed producers can maintain profitability while transitioning to sustainable alternatives:

- **High Costs of Novel Ingredients**: Many promising alternative protein sources, such as insect meals, single-cell proteins, and microalgae, are still in the early stages of commercial development. As a result, their production costs remain high compared to traditional feed ingredients. Scaling up production and improving processing efficiencies will be crucial to making these ingredients economically viable for widespread use in aquafeeds.
- **Access to Feed Ingredients in Developing Regions**: In many developing countries, where aquaculture is a vital source of food and income, access to affordable and nutritionally balanced feed ingredients is limited. The high cost of imported feed ingredients, coupled with the lack of local feed production infrastructure, presents significant barriers to the growth of aquaculture in these regions. Supporting local feed production and exploring regionally available feed ingredients, such as agricultural by-products and indigenous plant species, could help mitigate these challenges.

1. Challenges in Fisheries Nutrition

The field of fisheries nutrition faces numerous challenges as aquaculture continues to expand and diversify to meet global seafood demand. The industry must balance the nutritional needs of farmed fish, environmental sustainability, and economic viability. These challenges are intricately linked, as the reliance on traditional feed ingredients like fishmeal and fish oil is becoming unsustainable due to ecological, economic, and supply constraints. Below are key challenges that highlight the complexity of fisheries nutrition:

1.1. Dependence on Marine-Derived Ingredients

Fishmeal and fish oil have been the cornerstone of aquaculture feeds, particularly for carnivorous species such as salmon, trout, and sea bass. These marine-derived ingredients are highly valued due to their excellent nutritional profile, rich in essential amino acids, long-chain omega-3 fatty acids (EPA and DHA), and micronutrients vital for fish growth, immune function, and reproduction. However, the heavy dependence on fishmeal and fish oil poses several challenges:

- **Overfishing and Ecosystem Disruption**: The production of fishmeal and fish oil primarily depends on small pelagic fish species like anchovies, sardines, and menhaden. These species play crucial roles in marine ecosystems, serving as a food source for larger predators such as seabirds, marine mammals, and commercially valuable fish species. The increasing demand for fishmeal and fish oil has led to overfishing of these species, threatening marine biodiversity and disrupting food webs.
- **Supply Volatility**: The supply of marine-derived ingredients is subject to natural fluctuations driven by environmental conditions such as El Niño events, climate change, and changes in fish population dynamics. Regulatory measures, such as fishing quotas and marine conservation policies, also influence the availability of these resources. These fluctuations can cause price volatility, creating uncertainty for aquaculture producers who rely on stable feed costs to maintain profitability.
- **Rising Costs**: The depletion of wild fish stocks, coupled with growing demand from the expanding aquaculture industry, has driven up the prices of fishmeal and fish oil. This trend places financial pressure on fish farmers, especially in developing countries where access to cost-effective feeds is critical for sustaining production. The rising cost of these ingredients necessitates the exploration of alternative protein and lipid sources that are both nutritionally adequate and economically viable.

1.2. Nutritional Imbalance in Alternative Protein Sources

To address the challenges posed by fishmeal and fish oil reliance, numerous alternative protein sources have been explored, including plant-based proteins, insect meals, and microbial proteins. While these alternatives offer sustainability benefits, they present significant nutritional challenges:

- **Incomplete Amino Acid Profiles**: Many plant-based proteins, such as soybean meal, pea protein, and corn gluten, lack one or more essential amino acids that are critical for fish health and growth. For example, soybean meal is deficient in methionine, while pea protein lacks both methionine and cysteine. This imbalance requires the addition of synthetic amino acids or complementary ingredients, increasing formulation complexity and cost.
- **Anti-Nutritional Factors (ANFs)**: Plant-based proteins, especially soybean meal and rapeseed meal, contain ANFs such as protease inhibitors, phytic acid, and lectins, which can impair nutrient absorption, reduce feed efficiency, and negatively affect fish health. Although processing methods like fermentation and enzymatic treatments can reduce the impact of ANFs, these methods add to production costs and may not be feasible on a large scale.
- **Digestibility and Palatability Issues**: The digestibility of alternative protein sources varies across fish species, particularly for carnivorous fish that have evolved to consume high-protein, animal-based diets. Plant proteins often have lower digestibility than fishmeal, leading to reduced feed conversion ratios (FCR) and growth rates. Insects, while promising, can have variable digestibility depending on processing methods. Furthermore, palatability is· another concern, as some alternative proteins are less appealing to fish, leading to lower feed intake and growth performance.

1.3. Sustainability and Environmental Impact

Sustainability is a key driver of innovation in fisheries nutrition, yet aquafeed production itself can have significant environmental impacts. These impacts must be carefully managed to ensure the long-term viability of aquaculture as a sustainable food production system:

- **Environmental Footprint of Feed Ingredients**: The cultivation of plant-based protein sources like soybeans and corn requires large amounts of land, water, and agrochemicals. In regions such as the Amazon rainforest, soybean production has contributed to deforestation, biodiversity loss, and increased greenhouse gas emissions. Insect meals and microbial proteins, while more sustainable, face challenges in scaling up production to meet global demand, and the environmental impacts of their production processes are still being assessed.

- **Nutrient Pollution and Eutrophication**: One of the significant environmental challenges associated with aquaculture is nutrient pollution from uneaten feed and fish waste. High levels of nitrogen and phosphorus in effluents can lead to eutrophication of surrounding water bodies, causing harmful algal blooms and oxygen depletion, which can devastate aquatic ecosystems. Reducing nutrient losses from feeds, either through improved feed formulations or feeding practices, is essential to minimize the environmental impact of aquaculture.
- **Balancing Sustainability with Fish Health and Growth**: The need for sustainable feed ingredients must be balanced with maintaining fish health and productivity. Fish require nutritionally balanced diets to grow efficiently and resist disease. As alternative protein sources are integrated into aquafeeds, careful attention must be given to ensuring that these diets meet the specific nutritional needs of each fish species. Poorly formulated diets can compromise fish welfare, increase disease susceptibility, and reduce farm productivity, negating the environmental benefits of alternative feed ingredients.

1.4. Economic Viability and Access to Feed Ingredients

The economic challenge in fisheries nutrition is twofold: providing affordable feed ingredients for farmers and ensuring that feed producers can maintain profitability while transitioning to sustainable alternatives:

- **High Costs of Novel Ingredients**: Many promising alternative protein sources, such as insect meals, single-cell proteins, and microalgae, are still in the early stages of commercial development. As a result, their production costs remain high compared to traditional feed ingredients. Scaling up production and improving processing efficiencies will be crucial to making these ingredients economically viable for widespread use in aquafeeds.
- **Access to Feed Ingredients in Developing Regions**: In many developing countries, where aquaculture is a vital source of food and income, access to affordable and nutritionally balanced feed ingredients is limited. The high cost of imported feed ingredients, coupled with the lack of local feed production infrastructure, presents significant barriers to the growth of aquaculture in these regions. Supporting local feed production and exploring regionally available feed ingredients, such as agricultural

by-products and indigenous plant species, could help mitigate these challenges.

2. Opportunities for Innovation in Fisheries Nutrition

As aquaculture continues to expand, there are numerous opportunities to innovate and drive sustainability in fisheries nutrition. Key areas of focus include the development of alternative protein sources, the application of precision nutrition techniques, the use of functional feeds, and the adoption of circular economy principles. These innovations aim to reduce reliance on traditional marine-derived ingredients, enhance fish health and growth performance, and promote the sustainable use of resources.

2.1 Alternative Protein Sources and Sustainable Feed Ingredients

One of the most exciting avenues for innovation in fisheries nutrition is the development of alternative protein sources that can replace or supplement fishmeal and fish oil in aquafeeds. The growing awareness of the ecological impact of marine-derived ingredients has fueled research into plant-based, insect-based, and microbial protein alternatives.

- **Plant-Based Proteins**: Soybean meal, pea protein, and lupin meal are widely used plant-based alternatives to fishmeal. However, these ingredients often lack certain essential amino acids, such as methionine, and contain anti-nutritional factors that can affect nutrient absorption. Innovations in plant breeding and processing techniques, such as enzymatic treatment and fermentation, are helping to improve the nutritional quality and digestibility of plant proteins. For instance, fermented soybean meal has been shown to enhance the digestibility and nutrient availability for fish, making it a more viable alternative in aquafeeds.
- **Insect-Based Proteins**: Insects, particularly **black soldier fly larvae (BSFL)**, are emerging as a sustainable and nutrient-rich feed ingredient. BSFL meal contains high levels of protein, essential fatty acids, and minerals, making it an excellent substitute for fishmeal in aquafeeds. Studies have demonstrated that fish fed BSFL meal exhibit growth rates and feed conversion ratios (FCR) comparable to those fed fishmeal-based diets. Moreover, BSFL can be reared on organic waste, offering a circular approach to resource use and reducing the environmental footprint of feed production.
- **Microbial Proteins**: Microbial proteins, derived from single-cell organisms like yeast, bacteria, and algae, represent another promising alternative. For example, **single-cell algae** like **Spirulina** and **Chlorella**

are not only rich in protein but also provide valuable omega-3 fatty acids, vitamins, and antioxidants. Advances in cultivation technologies, such as photobioreactors and the use of genetic engineering to enhance protein yield, are making microbial proteins more cost-competitive. Additionally, **yeast-based proteins** offer the dual benefits of high protein content and immune-boosting properties, which can improve fish health and disease resistance.

- **Single-Cell Proteins (SCPs)**: Derived from bacteria, fungi, and algae, SCPs are rich in protein and essential nutrients. Advances in fermentation technologies are making SCPs a viable option for large-scale aquafeed production. For instance, the use of **methanotrophic bacteria**, which convert methane into protein, presents a novel way to produce high-quality feed ingredients from industrial by-products.

These alternative protein sources not only address the sustainability challenges of traditional feed ingredients but also offer opportunities to enhance fish growth, health, and feed efficiency.

2.2. Precision Nutrition and Personalized Feeding Strategies

Precision nutrition, which involves tailoring diets to the specific nutritional needs of individual fish species or even individual fish, represents a major opportunity for optimizing fish health, growth performance, and feed efficiency. Recent advances in genomics, proteomics, and metabolomics have paved the way for species-specific, and in some cases individual-specific, feed formulations.

- **Genomic Insights**: By understanding the genetic makeup of different fish species, researchers can identify specific dietary needs and metabolic responses to various nutrients. This allows for the development of optimized feeds that enhance growth and immune function while minimizing waste. For example, genomic studies have revealed variations in the ability of different fish species to utilize plant-based proteins, leading to more targeted formulations that improve feed conversion ratios.
- **Automated Feeding Systems**: Advances in **sensor technologies** and **artificial intelligence (AI)** have enabled the development of automated feeding systems that monitor individual fish behavior, health, and feeding patterns in real time. These systems can adjust feeding regimes based on the specific needs of individual fish, reducing feed wastage and ensuring that each fish receives the optimal amount of nutrition. AI-powered systems can also detect early signs of disease or stress,

allowing for timely interventions that improve fish welfare and reduce mortality.

- **Personalized Diets**: As precision nutrition techniques continue to evolve, there is potential to develop personalized diets for fish based on their genetic profile, health status, and environmental conditions. This approach could significantly improve growth performance, reduce feed costs, and enhance sustainability by minimizing nutrient excretion and waste.

2.3. Functional Feeds and Health-Promoting Additives

The use of functional feeds, which contain bioactive compounds that promote fish health, disease resistance, and growth performance, is another area ripe for innovation. Functional feeds not only meet the nutritional needs of fish but also enhance their overall well-being, reducing the need for antibiotics and chemical treatments.

- **Probiotics and Prebiotics**: Probiotics, which consist of beneficial bacteria, and prebiotics, which stimulate the growth of these bacteria in the gut, are increasingly being incorporated into aquafeeds to improve gut health and enhance immune function. These additives can help fish resist infections, reduce inflammation, and improve nutrient absorption, ultimately leading to better growth and survival rates.
- **Immunostimulants**: Nutritional immunostimulants, such as beta-glucans and nucleotides, have been shown to enhance the immune response of fish, making them more resilient to diseases and environmental stressors. These compounds can be derived from natural sources, such as yeast cell walls, and are increasingly being integrated into functional feeds to promote fish health and reduce reliance on antibiotics.
- **Phytogenic Compounds**: Plant-based compounds, including essential oils and herbal extracts from plants like garlic, ginger, and oregano, have gained attention for their antimicrobial, antioxidant, and anti-inflammatory properties. Phytogenics can improve fish growth performance, enhance feed palatability, and protect against bacterial and parasitic infections. Their natural origin also aligns with consumer demand for more sustainable and organic aquaculture practices.

2.4. Circular Economy and Waste Utilization

The adoption of a circular economy approach in aquaculture feed production presents significant opportunities for reducing waste, improving resource efficiency, and enhancing sustainability. By utilizing by-products from other industries and recycling waste from aquaculture operations, the environmental footprint of aquafeed production can be minimized.

- **Utilization of Industry By-Products**: The food processing and livestock industries generate large amounts of by-products, such as poultry offal, feather meal, and blood meal, which can be repurposed as high-protein feed ingredients for fish. These by-products are not only cost-effective but also reduce the need for fishmeal and other virgin raw materials, contributing to more sustainable feed production.
- **Aquaculture Waste Recycling**: Technologies such as **anaerobic digestion** and **composting** offer ways to recycle organic waste from aquaculture operations, including uneaten feed and fish excrement. Anaerobic digestion can convert organic waste into **biogas**, which can be used as a renewable energy source, while composting can produce nutrient-rich fertilizers that can be used in agriculture. Additionally, **biofloc systems**, which promote the growth of beneficial microbial communities in aquaculture ponds, can help recycle nutrients and improve water quality, reducing the need for external feed inputs.

3. Future Perspectives in Fisheries Nutrition

The future of fisheries nutrition is poised for transformative changes driven by technological advancements, collaborative efforts, and a deepening understanding of sustainable practices. As the demand for aquaculture products continues to surge, there is an urgent need to address the environmental challenges posed by traditional feed ingredients, such as fishmeal and fish oil. The following key areas represent promising directions for innovation and development in fisheries nutrition.

3.1 Novel Protein Sources

One of the most significant shifts in fisheries nutrition will be the exploration and utilization of novel protein sources, including genetically engineered microbes and synthetic proteins. These innovations have the potential to significantly reduce the environmental footprint of aquafeed production while providing the essential nutrients needed for the growth and health of farmed fish.

- **Genetically Engineered Microbes**: Advances in genetic engineering are enabling the development of microbes that can produce high-quality proteins tailored for aquaculture. These engineered organisms can utilize various substrates, including agricultural waste, to produce protein-rich biomass. For instance, researchers are exploring **yeast and bacteria** engineered to synthesize essential amino acids and omega-3 fatty acids, thereby providing complete nutritional profiles for fish diets. The scalability of microbial fermentation processes makes this approach particularly appealing for large-scale production.

- **Synthetic Proteins**: Synthetic biology is paving the way for the creation of lab-grown proteins that mimic the nutritional profiles of traditional feed ingredients. Companies are beginning to produce **synthetic fishmeal** through fermentation processes, which could serve as a sustainable alternative to fish-derived protein. This method not only minimizes overfishing but also reduces the associated greenhouse gas emissions and habitat destruction linked to conventional aquaculture practices.
- **Precision Fermentation**: This biotechnology utilizes engineered microbes to produce specific proteins in a controlled environment. The precision fermentation process allows for the customization of protein production to meet the precise nutritional requirements of different fish species. By optimizing conditions for microbial growth and protein synthesis, researchers can develop sustainable feed ingredients that cater to the diverse dietary needs of carnivorous fish.

3.2. Advances in Processing Technologies

Improving the digestibility and nutritional value of alternative protein sources is crucial for their successful incorporation into aquafeeds. Advances in processing technologies, such as enzymatic treatments and fermentation, will play a vital role in enhancing the quality of these ingredients.

- **Enzymatic Treatments:** Enzymes can be utilized to break down complex carbohydrates and proteins in plant-based feeds, improving their digestibility for fish. For example, using enzymes like proteases and carbohydrases can enhance the amino acid availability and overall nutrient profile of feed ingredients. This not only improves fish growth performance but also reduces nutrient excretion into the environment, addressing concerns related to water quality and eutrophication.
- **Fermentation:** The fermentation of alternative feed ingredients can significantly improve their nutritional value. Fermentation processes can increase the bioavailability of nutrients and reduce anti-nutritional factors present in raw materials. For instance, fermenting soybean meal can enhance its digestibility and nutritional quality, making it a more viable option for aquaculture diets. The fermentation process also produces beneficial metabolites that can support fish health and growth.
- **Microencapsulation:** This innovative technology allows for the controlled release of bioactive compounds and nutrients within fish feeds. By encapsulating sensitive ingredients, such as omega-3 fatty acids or probiotics, their stability and bioavailability can be enhanced, leading to improved health benefits for fish.

3.3. Policy Frameworks and Collaboration

For fisheries nutrition to evolve effectively, supportive policy frameworks and collaborative efforts among stakeholders are essential. Governments, academic institutions, and industry players must work together to create a conducive environment for innovation and research in aquaculture feed production.

- **Incentivizing Sustainable Practices**: Policies that promote sustainable feed practices can drive progress in the industry. Incentives for aquaculture producers to adopt alternative protein sources, utilize waste products, and implement precision nutrition strategies can help reduce reliance on traditional feed ingredients. Financial support for research and development initiatives can further encourage innovation in fisheries nutrition.
- **Public-Private Partnerships**: Collaborations between academia, industry, and government can facilitate knowledge transfer and accelerate the development of sustainable feed solutions. Joint research projects and pilot programs can help test and scale innovative feed technologies, providing valuable insights into their efficacy and commercial viability.
- **Regulatory Support**: Establishing regulatory frameworks that streamline the approval processes for novel feed ingredients and production technologies can help bring innovative solutions to market more rapidly. Ensuring that these regulations are science-based and promote safety and efficacy will be crucial for maintaining consumer confidence in aquaculture products.

Conclusion

Fisheries nutrition is at a critical juncture, where the challenges posed by traditional feed ingredients present both significant obstacles and exciting opportunities for innovation. As the aquaculture industry seeks to meet the increasing global demand for seafood, it must embrace alternative protein sources, leverage advances in processing technologies, and foster collaborative efforts among stakeholders. By prioritizing sustainable practices and investing in research and development, the aquaculture sector can move toward a more resilient and environmentally responsible future. Innovations such as genetically engineered microbes, precision fermentation, and enhanced processing techniques hold the promise of delivering high-quality feed ingredients that meet the nutritional needs of farmed fish while minimizing the ecological impact of aquaculture. The ongoing evolution of fisheries nutrition will not only contribute to food security but also ensure the health of marine ecosystems for generations to come.

References

Ali, M. A., & A. K. (2021). The Future of Aquaculture: An Integrated Approach to Fisheries Nutrition. Fisheries Research, 233, 105761.

Ali, M. R., & A. Z. (2022). Emerging Trends in Fisheries Nutrition: Challenges and Innovations. Marine Policy, 138, 104975.

Barrows, F. T., & C. H. (2019). Emerging Aquaculture Feeds: A Review of Alternative Protein Sources. Reviews in Fisheries Science & Aquaculture, 27(3), 327-343.

Birk, S., & G. H. (2019). Nutritional Composition of Insects and Their Potential in Aquaculture. Insects, 10(1), 1-20.

Brown, P., & K. R. (2018). Nutritional Strategies for Enhancing Fish Health and Welfare. Aquaculture Reports, 10, 59-66.

Doyon, J. F., & R. S. (2020). Aquaculture and the Circular Economy: Opportunities for Innovation. Aquaculture Reports, 17, 100363.

FAO (2020). The State of World Fisheries and Aquaculture 2020. Rome: FAO.

Glencross, B. D., & S. A. (2016). Alternatives to Fishmeal and Fish Oil in Aquaculture Feeds. Aquaculture Nutrition, 22(4), 870-887.

Gomes, E. F., & F. C. (2017). Fishmeal Alternatives in Aquaculture: The Role of Insects. Sustainability, 9(12), 2240.

Hertrampf, J. W., & P. L. (2012). Fish Nutrition in Aquaculture: An Overview of Current Practices. Aquaculture Research, 43(7), 1091-1100.

Kauffman, J. F., & G. R. (2019). Utilizing Food Processing By-products in Aquaculture Feeds. Journal of Cleaner Production, 229, 614-622.

Khosravi, F., & A. F. (2021). Sustainability Assessment of Aquaculture Feed Ingredients. Aquaculture Research, 52(5), 1744-1760.

Lee, S. M., & H. J. (2019). Sustainable Protein Sources for Aquaculture. Aquaculture Research, 50(8), 2326-2338.

Liti, D., & G. K. (2020). The Role of Policy in Advancing Sustainable Aquaculture. Aquaculture Economics & Management, 24(3), 262-276.

Naylor, R. L., & D. R. (2021). The Future of Aquaculture: Integrating Sustainability into Fisheries Nutrition. Nature Sustainability, 4(7), 655-665.

Naylor, R. L., & G. M. (2020). A Global Perspective on Sustainable Aquaculture. Nature Sustainability, 3(6), 432-434.

O'Leary, B. C., & J. S. (2019). Advancements in Aquaculture Feed Technology and Sustainability. Environmental Science & Policy, 94, 91-97.

O'Neill, J., & D. H. (2020). Environmental Impacts of Fish Feed Ingredients. Aquaculture Environment Interactions, 12, 153-169.

Ochoa, F., & G. F. (2020). The Role of Biotechnology in Sustainable Aquaculture Feed Production. Aquaculture Nutrition, 26(3), 1103-1114.

Riche, M., & G. R. (2018). Aquaculture Nutrition: A Practical Guide to Feed Formulation and Fish Health. Wiley-Blackwell.

Roque, A., & R. S. (2019). Functional Feeds: A Novel Approach in Aquaculture. Aquaculture, 509, 85-90.

Sarker, P. K., & S. S. (2021). Functional Ingredients for Sustainable Aquaculture Feeds. Aquaculture Research, 52(12), 5983-6001.

Sprague, M., & S. S. (2019). Microalgae as a Sustainable Feed Ingredient for Aquaculture. Aquaculture Reports, 13, 14-19.

Stoll, A., & B. H. (2020). Precision Nutrition in Aquaculture: A Review. Aquaculture International, 28(4), 1201-1221.

Tacon, A. G. J., & Metian, M. (2013). Fish Matters: Importance of Aquatic Foods in Human Nutrition and Global Food Security. Food Security, 5(5), 509-521.

Wan, X., & Z. Y. (2020). Insect Meal as an Alternative Protein Source for Fish: A Review. Fish Physiology and Biochemistry, 46(2), 481-494.

Wang, Y., & J. J. (2021). Innovations in Fish Feed Formulation: The Future of Fisheries Nutrition. Aquaculture Nutrition, 27(6), 1300-1312.

Xu, D. P., & H. Z. (2020). Innovations in Fish Feed Technology: Trends and Future Directions. Trends in Food Science & Technology, 99, 22-29.

Ytrestøyl, T., & A. T. (2015). Sustainable Use of Plant Protein Sources in Aquaculture. Reviews in Fisheries Science & Aquaculture, 23(4), 325-339.

Zheng, K., & Z. R. (2018). Development of Alternative Protein Sources for Aquaculture. Journal of Aquaculture Research & Development, 9(2), 1-10.

Index

M

N

O

P

S

T

V

W